Birkhäuser

Trends in the History of Science

Trends in the History of Science is a series devoted to the publication of volumes arising from workshops and conferences in all areas of current research in the history of science, primarily with a focus on the history of mathematics, physics, and their applications. Its aim is to make current developments available to the community as rapidly as possible without compromising quality, and to archive those developments for reference purposes. Proposals for volumes can be submitted using the online book project submission form at our website www.birkhauser-science.com.

More information about this series at http://www.springer.com/series/11668

Shier Ju • Benedikt Löwe • Thomas Müller •
Yun Xie
Editors

Cultures of Mathematics and Logic

Selected Papers from the Conference in Guangzhou, China, November 9-12, 2012

Birkhäuser

Editors
Shier Ju
Institute of Logic and Cognition
Sun Yat-Sen University
Guangzhou, P.R. China

Benedikt Löwe
Fachbereich Mathematik
Universität Hamburg
Hamburg, Germany

Institute for Logic, Language
and Computation
Universiteit van Amsterdam
Amsterdam, The Netherlands

Thomas Müller
Fachbereich Philosophie
Universität Konstanz
Konstanz, Germany

Yun Xie
Institute of Logic and Cognition
Sun Yat-sen University
Guangzhou, P.R. China

ISSN 2297-2951 ISSN 2297-296X (electronic)
Trends in the History of Science
ISBN 978-3-319-31500-3 ISBN 978-3-319-31502-7 (eBook)
DOI 10.1007/978-3-319-31502-7

Library of Congress Control Number: 2016947424

Printed on acid-free paper

This book is published under the trade name Birkhäuser
The registered company is Springer International Publishing AG Switzerland

Preface

Mathematics and formal reasoning are fundamental building blocks of knowledge, essential for science, technology, policy-making, and risk-management. Mathematical practice is a rich phenomenon of human activity, with subtle differences between various cultures. Here, the word "culture" can refer to national cultures, but also cultural differences in different historical periods, in different strata of a given society, or in different social settings.

And yet, the public perception of mathematics is that of an apersonal subject with little or no human interaction, based on a false picture of a science of pure thought and deduction, with almost no interaction or visible activity.

In a move away from these traditionalist positions, philosophers and social scientists have recently become more interested in studying mathematical and logical practice, or more accurately, the differences between mathematical and logical practices. The conference *Cultures of Mathematics and Logic* held in Guangzhou, China, was the third in a series of interdisciplinary, international conferences that strive to bring together researchers from different fields and different backgrounds for in-depth discussions about the role and impact of culture(s) on the practice of the formal sciences of mathematics and logic. The community meeting at these conferences consists of philosophers of mathematics, historians of mathematics, sociologists of mathematics, anthropologists, cognitive scientists, and researchers in mathematics education. Previous meetings of this series were *Mathematics as Culture and Practice*, held in Bielefeld, Germany (May 2010), and *Mathematics as Culture and Practice II*, held in Greifswald, Germany (December 2011). While members of our community believe that an ambient (non-mathematical) culture affects mathematical practices, so that there are differences between national styles of mathematics, these differences can be difficult to observe and isolate, especially when the ambient cultures are close to each other. It could be expected that these differences become more pronounced if the ambient cultures differ more substantially; given the dominance of Western cultural influences on mathematics and logic, the organisers of this conference felt that it would be appropriate to have the next meeting in a country with a strong and different ambient cultural tradition such as China. Naturally, discussions of Chinese mathematics became an important topic at the conference. Based on the same idea, the follow-up conference *Cultures of Mathematics IV* took place on 22–25 March 2015 in New Delhi, India.

The Guangzhou conference took place from 9 to 12 November 2012, at the campus of one of the sponsoring institutions, Sun Yat-Sen University. It was organized jointly by Shier Ju (Guangzhou), Benedikt Löwe (Amsterdam and Hamburg), Thomas Müller (then Utrecht, now Konstanz), and Yun Xie (Guangzhou). The programme committee consisted of Mihir Chakraborty (Kolkata), Shuchun Guo (Beijing), Joachim Kurtz (Heidelberg), Brendan Larvor (Hatfield), Benedikt Löwe, Martina Merz (Luzern), Thomas Müller, Dirk Schlimm (Montréal QC), and Shier Ju.

For the multi-disciplinary community behind these meetings, the notion of "practices and cultures" is at the same time central and underdetermined, and two of the papers in this volume can be seen as a discussion of our community and its aims. The article *What are mathematical cultures?* by Brendan Larvor serves as a discussion of what we might mean by *cultures* in this context; at the same time, it provides an overview of the activities and publications of our community during the last decade, which to a large extent happened under the umbrella of the *Philosophy of Mathematical Practice* movement. That the discipline of philosophy plays such an important role for our community can come as a surprise to some; the scope of our field is explored in Benedikt Löwe's *Philosophy or not? The study of cultures and practices of mathematics* where he argues that our field forms a multi-disciplinary community and discusses the role of philosophy for this community as a whole. The reader is invited to start with the papers by Larvor and Löwe to get an overview of the overall aims and scope of the community that is behind the Guangzhou conference. The talks at the Guangzhou conference took place within the framework described in these two papers. The following list documents all talks invited or accepted for presentation at our conference, including those that had to be cancelled due to various reasons:

Invited Speakers

Andrea Bender. Numeration systems as cultural tools
Karine Chemla. Practices of abstraction as features of a mathematical culture
Shirong Guo. The reasoning and its logical structure in traditional Chinese mathematics
Juan Pablo Mejía Ramos. Reading mathematics: Empirical research on expert mathematical practice
Reviel Netz. Mathematical communities in Greek antiquity
Zhaoshi Zeng and Gang Wang. Study of Chinese logic from the perspective of the General Argumentation Theory

Contributed talks

Mihir Chakraborty and Smita Sirker. Some aspects of mathematical pluralism
Amita Chatterjee. Logical subcultures in the Classical Indian theoretical tradition
Karen François. The cultural turn within the research field of mathematics education

Yang He and Yanjin Chen. On Gongsun Long's methods of argumentation
Albrecht Heeffer. From tables to induction in Abbaco mathematical culture
Peter Koepke. Formal mathematics and mathematical practices
Joachim Kurtz. De-modernizing the history of Chinese logic
Brendan Larvor. The *Mathematical Cultures* research network
Baptiste Mélès. Programming languages for pre-mechanical calculating tools
Thomas Müller. Is there such a thing as philosophical logic?
Ranjit Nair. Philosophies of mathematics, logic and language: East and West
Markus Pantsar. Philosophy of mathematics in different fields
Stig Andur Pedersen. Mathematics in engineering and science
Josipa Petrunic. Revolutions and epistemic cultures: The case of Hamilton's quaternions as an epistemic shift and a mathematical revolution
Mario Piazza, Gabriele Pulcini, and Nevia Dolcini. Patterns of mathematical cognition: The prototypical proof
Dagmar Provijn. Much reasoning, many logics: on dynamics and heuristics in reasoning
Johannes Wietzke. The desire for knowledge in the Greek exact sciences
Jia-Ming Ying. The style of argumentation in emperor Kangxi's mathematical compendium and its influence on Korean mathematics
Yijie Zhang. Liu Hui's inference in *The Nine Chapters on the Mathematical Procedures*: A preliminary inquiry
Dahai Zou. The foundations of the reasoning in the demonstration of Liu Hui's principle

The conference was funded jointly by the Institute of Logic and Cognition, Sun Yat-Sen University, China, the Department of Philosophy at Utrecht University, The Netherlands, and the Institute for Logic, Language and Computation at the University of Amsterdam, The Netherlands. We should like to thank especially the local organizational team at Guangzhou, led by Ju Shier and Yun Xie, who provided a welcoming atmosphere and an efficient local organization. We are also grateful for the work of our programme committee and for the diligent work of the many referees that helped with the time-consuming task of selecting the best among submitted papers. The editors insisted on the highest standards of journal refereeing for all papers. They accepted only those that met the criterion of scientific excellence in order to produce a high-quality volume that reflects the topics discussed at the conference.

December 2015 S.J., B.L., T.M., Y.X.

Contents

What Are Mathematical Cultures?

Brendan Larvor

Abstract

In this paper, I will argue for two claims. First, there is no commonly agreed, unproblematic conception of culture for students of mathematical practices to use. Rather, there are many imperfect candidates. One reason for this diversity is there is a tension between the material and ideal aspects of culture that different conceptions manage in different ways. Second, normativity is unavoidable, even in those studies that attempt to use resolutely descriptive, value-neutral conceptions of culture. This is because our interest as researchers into mathematical practices is in the study of successful mathematical practices (or, in the case of mathematical education, practices that ought to be successful).

I first distinguish normative conceptions of culture from descriptive or scientific conceptions. Having suggested that this distinction is in general unstable, I then consider the special case of mathematics. I take a cursory overview of the field of study of mathematical cultures, and suggest that it is less well developed than the number of books and conferences with the word 'culture' in their titles might suggest. Finally, I turn to two theorists of culture whose models have gained some traction in mathematics education: Gert Hofstede and Alan Bishop. Analysis of these two models corroborates (in so far as two instances can) the general claims of this paper that there is no escaping normativity in this field,

I am grateful to Paul Ernest, as well as to two anonymous referees and the editor of this collection for comments on drafts of this paper. I am also grateful to the organisers of *Cultures of Mathematics and Logic* (9–12 November 2012, Institute for Logic and Cognition, Sun Yat-Sen University, Guangzhou, China)

B. Larvor (✉)
University of Hertfordshire, College Lane, Hatfield, AL10 9AB, UK
e-mail: b.p.larvor@herts.ac.uk

S. Ju et al. (eds.), *Cultures of Mathematics and Logic*, Trends in the History of Science, DOI 10.1007/978-3-319-31502-7_1

and that there is no unproblematic conception of culture available for students of mathematical practices to use.

1 The Normative Senses of 'Culture'

The word 'culture' is semantically rich. Its many meanings divide into two broad groups: normative/educative and descriptive/scientific. On one side, there are the normative senses in which culture is a good thing, valuable in its own right and for the people who have it, both individually and collectively. Thus, a person may be *cultured.* Such use of the term 'culture' requires a high level of confidence in the associated evaluations. One must firmly believe that one knows what is valuable in order to deploy the word in this way, and be confident of culture's good effects. For this reason, in Europe, this normative sense of culture finds its clearest expression in high-minded men of the nineteenth century, such as Matthew Arnold.

According to Arnold, writing in 1869, culture recommends itself as

> ...the great help out of our present difficulties; culture being a pursuit of our total perfection by means of getting to know, on all the matters which most concern us, the best which has been thought and said in the world, and, through this knowledge, turning a stream of fresh and free thought upon our stock notions and habits...[1]

Among the 'present difficulties' that Arnold hoped to resolve with the help of culture were the uncouth, unruly and sometimes violent behaviour of the recently formed urban proletariat and the mercantile values of the proprietors of small businesses who funded the schools that he inspected. For Arnold, culture is an educational project. Everyone can and should have it; everyone can and should benefit from it. While his vision retains much of its attraction, his confidence that a curriculum committee can know what is 'the best which has been thought and said in the world' must strike us as naïve. While a present-day curriculum committee might be confident that what it lays before students is good, it cannot know that there is not, somewhere, a better thought, perhaps expressed in a language that none of the committee members understands. Moreover, Arnold's confidence that culture can be 'the great help out of our present difficulties' is hard to share now. Most developed countries have had compulsory universal education for most of the past century, and this normally includes plenty of culture as Arnold understood it. Four generations of British children have all had their dose of Shakespeare, for example. Perhaps this has helped to make Britain less violent and money-grubbing than it would otherwise have been—counterfactual history on that scale is an uncertain exercise. What is clear is that the 'present difficulties' that troubled Arnold are still with us in varying degrees.

[1] *Culture and Anarchy*, preface (Cambridge University Press 1978 reprint of the second (1875) edition, p. 6). Also of interest is Schiller (1794), because he attempted to explain in detail how exposure to high art can do its ennobling work.

Nevertheless, culture, thus understood, is a good and nurturing thing. It stands in contrast to nature and ennobles the youth fortunate to be educated in it. Culture in this sense was usually understood to be what we would now call 'high culture'—that is, canonically great works of high art. Even Arnold did not expect folk-tales to help us with our present difficulties, unless a literary author turns the tales into high art.

Since Arnold's day, it has become a truism that highly cultured people can commit terrible crimes. Nevertheless, some version of Arnold's view survives in every educator who insists that exposure to high culture can render the cultured individual wiser and more virtuous as well as wittier and better informed.

Here then is a question for mathematics educators: do we subscribe to a mathematical version of Arnold's view of culture? Given that few people now need any mathematical knowledge or skill beyond elementary arithmetic, why is mathematics compulsory for all school pupils? There are, of course, some answers to this question that do not invoke any version of Arnold's view of culture. Citizenship and prudence require some understanding of statistical and actuarial reasoning, and besides, it may not be possible to know exactly which children will grow up to be the minority who do use non-elementary mathematics, so it may be best to apply a sheep-dip approach and train them all. Still, we should consider the possibility of giving an answer in the style of Matthew Arnold, that is, that (a) mathematics is good in itself and therefore the person who appreciates it is richer in spirit and (b) appreciation of mathematics has some general educative effect, beyond mastery of the knowledge and skills that constitute it.[2] This reflection could go either way: we might decide to work out in detail a view like Arnold's for mathematics, or we might decide that it is indefensible and take care to expunge every trace of it from our doctrine.

For the moment, it almost suffices to distinguish the normative/educative senses of 'culture' from the descriptive/scientific senses, as the latter will occupy most of the remainder of this essay. Almost, because every distinction establishes a relation, and echoes of the normative/educative sense are detectable throughout the descriptive and scientific literature on culture in general and mathematical cultures in particular. Some cultural approaches to mathematics, such as the Critical Mathematics Education movement associated with Ole Skovsmose, show this interplay

[2]There is, for example, more than a trace of Matthew Arnold in Paul Lockhart's *Mathematician's Lament* (Bellevue Literary Press, 2009). Paul Ernest makes an Arnold-like argument in 'Why Teach Mathematics?' (in White and Bramall (2000) *Why Learn Maths?* London: Institute of Education, 2000), including a distinction between capability and appreciation that emphasises the value of mathematics as cultural achievement. In the closing remarks of his presentation at the third London conference on mathematical cultures (2014), Ernest advocated inspiring pupils with "the poetry of mathematics".

between the normative and descriptive senses of 'culture' explicitly.[3] However, we should expect to see a normative notion of culture implicit in any culturally-focussed research directed at the question 'How can we teach mathematics *better*', even if that research makes explicit use of descriptive or scientific concepts of culture.

2 The Descriptive/Scientific Senses of 'Culture'

On the descriptive, scientific or anthropological sense of the term 'culture', a good place to start is the magisterial literature review in Kroeber, A.L., & Kluckhohn, C. (1952) *Culture: A critical review of concepts and definitions.*[4] This report starts with the first recorded scholarly use of the term and tracks its evolution and differentiation up to their time of writing. This is of some interest, given the discussion so far, because Kroeber and Kluckhohn explore both sides of the descriptive/normative distinction, and their historical survey includes those German thinkers of the nineteenth century who distinguished 'ought' from 'is' but took care not to raise this distinction into an absolute separation.[5]

At the end of their historical journey, they land on this definition of culture, which they take to be an approximation of the view of "most social scientists":

> Culture consists of patterns, explicit and implicit, of and for behavior acquired and transmitted by symbols, constituting the distinctive achievements of human groups, including their embodiments in artifacts; the essential core of culture consists of traditional (i.e., historically derived and selected) ideas and especially their attached values; culture systems may, on the one hand, be considered as products of action, and on the other as conditioning elements of further action.[6]

In this definition, we see a tension that turns up more widely in the descriptive senses of culture. The first part of this definition presents culture as patterns of behaviour, embodied in and transmitted by symbols and artefacts. From a positivist or behaviourist point of view, that sounds scientifically respectable. Symbols, artefacts and behaviours are all empirically detectable (setting aside the question of how a purely empirical consciousness could recognise them *as* symbols, artefacts and behaviours). However, they go on to say that the core of (a) culture is "ideas and especially their attached values". Ideas and values are not so easy to detect empirically, except perhaps indirectly through their effects. This mention of values

[3]See Skovsmose's principal programmatic work *Towards a Philosophy of Critical Mathematics Education* (1994) Dordrecht: Kluwer. For his more recent reflections, see the interview in Alrø, Ravn and Valero (eds) *Critical Mathematics Education: Past, Present and Future* pp. 1–9 (2010), in which he relates his thinking to philosophers associated with critical theory such as Habermas, Adorno and Foucault. Naturally, his relation to these figures is not uncritical. From such perspectives, the present argument (that our interests in mathematics and mathematics education must inevitably erode the is/ought distinction) will appear as a naïve statement of the obvious.

[4]Harvard University Peabody Museum of American Archeology and Ethnology Papers 47.

[5]On the is/ought relation, see Hegel (1821) *Philosophy of Right.*

[6]Kroeber & Kluckhohn 1952 p. 181.

is not an afterthought; on the contrary, for Kroeber & Kluckhohn, values are central to anthropology: "Values provide the only basis for the fully intelligible comprehension of culture, because the actual organisation of all cultures is primarily in terms of their values."[7] It is also worth noting their use of the term 'achievement'. This is a success-word, and therefore invokes a normative sense of culture, though presumably the criteria of success are culture-specific.

The conceptions of culture that social anthropologists have developed since Kroeber & Kluckhohn collectively reflect this tension between empiricism (attending to empirically detectable features such as behaviours and artefacts) and idealism (focussing on ideas and values) in social science. In an essay published shortly before Kroeber & Kluckhohn's review, Talcott Parsons defined culture as "...those patterns relative to behavior and the products of human action which may be inherited, that is, passed on from generation to generation independently of the biological genes".[8] It may be significant, given that students of mathematical cultures are often researchers in mathematics education, that Parsons makes non-biological transmission definitive of culture. In any case, he focuses on behaviour and its products, and in doing so represents one of the main trends of mid-century social science. In contrast, writing in 1989, Banks, Banks, & McGee claim that,

> Most social scientists today view culture as consisting primarily of the symbolic, ideational, and intangible aspects of human societies. The essence of a culture is not its artifacts, tools, or other tangible cultural elements but how the members of the group interpret, use, and perceive them. It is the values, symbols, interpretations, and perspectives that distinguish one people from another in modernized societies; it is not material objects and other tangible aspects of human societies. People within a culture usually interpret the meaning of symbols, artifacts, and behaviors in the same or in similar ways.[9]

If this is indeed how most social scientists viewed culture in 1989, then it seems that attention has shifted away from behaviours and material artefacts, towards the ideas and values that Kroeber & Kluckhohn insisted are the only basis for understanding cultures.

We should expect to find this taut duality in any study of culture because the two ends need each other. The behaviours and artefacts do not classify themselves or explain themselves; the ideas and values must body forth in words, deeds and things if they are to have any presence at all. We can find this tension in the philosophy of mathematical practice. Mathematics is obviously concerned with ideas, but studying mathematical practices directs attention to artefacts (blackboards, notations, diagrams, models and computers) and behaviours (gesticulating, writing, sketching, gathering mathematicians in groups of various sizes, etc.). Note the reference to

---

[7]Kroeber & Kluckhohn 1952 p. 340.

[8]Parsons, Talcott (1949) *Essays in Sociological Theory.* Glencoe, IL, p. 8. Similarly, Useem & Useem define culture as, "... the learned and shared behavior of a community of interacting human beings" (Useem, J., & Useem, R. 1963 p. 169). That is, culture is patterned behaviour rather than ideas and values, and is reproduced non-biologically.

[9]Banks, J.A., Banks, & McGee, C. A. (1989). *Multicultural education.* Needham Heights, MA: Allyn & Bacon.

'modernized societies'. Presumably, the point there is that the material cultures and outward behaviours of such societies are increasingly homogenous—people all over the world wear jeans and use smart-phones. In order to study the differences between such societies, anthropologists have to pass from artefacts and behaviours to ideas and values. This thought may also have some use in the study of mathematical cultures.

3 Mathematics

There are two intellectual movements of recent decades that study mathematical cultures: philosophy of mathematical practice (which, with some clumsiness and injustice, we take to include corresponding changes in the historiography of mathematics) and the cultural turn in mathematics education research.

The philosophy of mathematical practice movement arises from a dissatisfaction among philosophers and historians with abstract models of mathematics that make a mystery of its growth and fail to explain how finite, embodied, naturally evolved creatures can understand it. This movement has now acquired a scholarly literature,[10] regular conferences (see Appendix for a partial list) and an international society, the Association for the Philosophy of Mathematical Practice (APMP).[11]

So far, the philosophy of mathematical practice movement has not done much to develop cultural approaches to mathematics. The number of meetings with the word 'culture' in the title is misleading; in many instances, the name of the meeting represents an aspiration on the part of the organisers rather than an accurate indication of the eventual contents. For the most part, philosophers and historians of mathematical practice have tried to study what mathematicians do and how these activities produce mathematics without invoking culture or related notions such as society or intersubjectivity. Notably, cultural approaches are largely absent from the most important book in the field in recent years, *The Philosophy of Mathematical Practice*, edited by Paolo Mancosu, and in this respect the papers gathered in this book are representative of the field.[12] Properly cultural questions about the social embedding and reproduction of practices have not yet had much attention (for example, we are still waiting for a systematic study of the expressions and enforcement of status and hierarchy in mathematics seminars).

[10]See Larvor's review of *The Philosophy of Mathematical Practice* Paolo Mancosu (ed.) OUP 2008 in *Philosophia Mathematica* (2010) 18(3): 350–360 for a representative list.

[11]http://www.philmathpractice.org/.

[12]Oxford University Press, 2008.

There have been sporadic and largely programmatic efforts; one of the earliest was the work of the topologist Raymond L. Wilder.[13] Beginning in 1950, Wilder proposed that mathematics is, essentially, culture. Perhaps because he offered the cultural approach as an answer to the metaphysical question 'what is mathematics?', Wilder's work did not directly stimulate a stream of scholars to take up his ideas. It may also have suffered from his relatively under-theorised notion of culture. The many insights in Wilder's main exposition of his view, *Mathematics as a Cultural System,*[14] seem to originate in his experience as a creative mathematician, rather than in the deployment of anthropological theory. The cultural approach seems to have liberated him to express general thoughts about mathematics that he believed on other grounds. This is valuable, but it is not a programme of research that others can take up and continue—unless they too are professional mathematicians with insights grounded in their personal experience of mathematical research. Wilder may have influenced some others, such as Phillip J. Davis, Reuben Hersh, David Bloor and perhaps Alvin White, who founded and for many years edited (what is now called) the *Journal of Humanistic Mathematics*. Hersh included some cultural or anthropological approaches in his collection *18 Unconventional Essays on the Nature of Mathematics* (Springer, 2006). One is by Leslie White, who introduced Wilder to anthropology; others are by Andrew Pickering, Eduard Glas and Hersh himself. As Hersh notes in his introduction (p. xv), most of the articles in this collection are on the cognitive aspects of mathematical practice rather than the cultural or social aspects, and in this it reflects the state of the field.

In recent years, the philosophy of mathematical practice movement has begun to show more interest in cultural approaches to mathematics, in the form of conferences, including the one that produced this book (Cultures of Logic and Mathematics, held in Guangzhou, China, November 2012). In Europe, the PhiMSAMP (Philosophy of Mathematics: Sociological Aspects and Mathematical Practice) project sought to connect sociologists of mathematics with historians and philosophers. In 2012–2014, London was the venue for a series of three conferences on mathematical cultures.[15] From the presentations in this series, it is clear that the concept of culture is still relatively untheorised among writers on mathematical practice. Only a few of the philosophers and historians at the meetings deployed a notion of culture explicitly, and these were mostly lacking in conceptual articulation compared with the concepts of culture developed by Kroeber & Kluckhohn and the other social scientists mentioned earlier in this essay. One notable exception is Albrecht Heeffer, whose presentation on the Abbaco mathematical culture made

[13]Wilder offered this definition of culture: "We use [the term 'culture'] in the general anthropological sense . . . In this sense, a culture is the collection of customs, beliefs, rituals, tools, traditions, etc., of a group of people . . . It is not the use of the term as in "a cultured person" that we have in mind." *Introduction to the Foundations of Mathematics* (John Wiley; second ed. 1965 (first published 1952)) p. 282.

[14]Pergamon Press, 1981.

[15]For a description of the series, see "The Mathematical Cultures Network Project" 2012 *Journal of Humanistic Mathematics* 2 (2): 157–160.

careful use of some of Alan Bishop's ideas on mathematical enculturation (on which more below). The dearth of cultural theory in this area is a serious deficiency, because practices, including mathematical practices, do not stabilise and reproduce themselves by magic. They have to be culturally embedded, manifested and valued. For this observation to motivate meaningful research, we need a notion of culture with some explanatory potential.

A point of particular interest is the cultural stability and reproduction of proof-practices, since this is the point where cultural studies meet epistemology. A central claim of the philosophy of mathematical practice movement is that formal logic is a poor model of mathematical proof, that 'real' mathematical proofs employ specifically mathematical means of inference rather than the topic-neutral inference-patterns modelled in formal logic.[16] Now, ideally, one would hope that adepts in any epistemic practice would be able to say why it works. One can think of mathematical logic as an attempt by mathematicians to achieve this ideal (though the history of mathematical logic does not entirely bear this out, and current research in formal logic mostly has other motivations). In any case, one of the motivations of the philosophy of mathematical practice movement is dissatisfaction with formal logic as an account of how and why mathematical proofs work. In place of a universal account—that mathematicians everywhere and at all times employ, however obscurely and informally, a common, topic-neutral reasoning ability—scholars in this movement look at local, topic-specific proof-practices. Given the centrality of ancient Greek mathematics in the history of western mathematics, it is not surprising that two of the best developed studies are in this tradition: Kenneth Manders' (2008) 'The Euclidean Diagram'[17] and Reviel Netz's *The Shaping of Deduction in Greek Mathematics.*[18] These studies lay out in detail the norms governing ancient Greek mathematics and attempt to show how these are grounded in practices. Netz (who spoke at the Guangzhou conference in 2012) claims that, "What unites a scientific community need not be a set of *beliefs*. Shared beliefs are much less common than shared practices . . . because shared beliefs require shared practices, but not vice versa." (1999, p. 2). Since the practices that (on this view) unite a scientific community are not normally the objects of its enquiries, we should not expect scientific communities to be able to explain their epistemic practices all the way down. Moreover, we should expect these practices to change—Netz is explicit about this, subtitling his book a 'study in cognitive history'. Rules of reasoning, he says (p. 6), are strictly historical. In that case, the question of stability and reproduction is especially acute in the case of mathematics, because some mathematical norms seem to remain stable over long periods, through dramatic social, political and religious changes, sometimes with gaps where the practices lay latent in documents, and in spite of the relevant community being quite

[16]This claim is argued many times over in the literature; something like a locus classicus is Rav's 'Why do we prove theorems?' *Philosophia Mathematica* (1999) 7 (1): 5–41.

[17]In Mancosu 2008

[18]Cambridge University Press, 1999.

small.[19] (Jody Azzouni develops this point in his (2006) contribution to Hersh's *18 Unconventional Essays*). Netz mentions the *Annales* historians' emphasis on the "conservation of practice in the material domain" (p. 4). (The *Annales* group of historians emphasised historical studies of long periods.) Nevertheless, the contrast between volatile human culture and enduring mathematical achievement is a standing problem for a cultural approach to mathematics.

A notable feature of the philosophy of mathematical practice movement is the paucity of contact and overlap with the sociology of mathematics, at least in the English-speaking world. This may in part have a sociological explanation, in that the engagement or lack of it between sociologists and members of the mathematical practice movement is partly conditioned by how these academic formations locate each other with respect to a constructed mainstream. In Hersh's *18 Unconventional Essays*, sociologists and practice-oriented philosophers gather as fellow mavericks. The sociologists Donald MacKenzie and Andrew Pickering sit in the same volume as the philosophers Jody Azzouni and Andrew Aberdein, all positioned outside an implied mainstream that writes conventional essays on mathematics. On the other hand, David Bloor's 'Strong Sociological Programme' was supposed to oppose and supplant normative philosophy of science. In the 'science wars' of the 1990s, philosophers and sociologists occupied opposing trenches. While we have to hope that peace is now permanent, there is as yet little trade between the former combatants. This may be because the origins of the mathematical practice movement in dissatisfaction with formal logic *as logic* (that is, as the normative study of inference) gives its members a different interest from the sociologists. Certainly, this is how Netz distances himself from sociologists whom he otherwise respects,

> I do not ask just what made science the way it was. I ask what made science successful, and successful in a real intellectual sense. In particular, I do not see 'deduction' as a sociological construct. I see it as an objectively valid form . . . [20]

Similarly, Manders' analysis in 'The Euclidean Diagram'[21] sets out to explain the success of ancient Greek geometry. In his preamble, he observes that "Euclid . . . Apollonius and Archimedes, are virtually without error".[22] In contrast, Claude Rosental's *Weaving Self-Evidence: A Sociology of Logic* tracks the emergence of a new theorem without making any reference to the validity of its proof. In a broad sense of 'logic', Netz and Manders are logicians, but Rosental is not.

[19]Netz makes this observation, ". . . mathematics survived even Christianity, and in the totally dissimilar culture of Islam *the very same* mathematics went on." (p. 237, emphasis added).

[20]1999, p. 3.

[21]Mancosu 2008, pp. 80–133.

[22]Mancosu 2008, p. 67.

While the philosophy of mathematical practice community has been relatively slow to take up cultural approaches to mathematics, there has been a turn towards culture in mathematics education research. Karen Francois, in her video presentation to the cultures of logic and mathematics conference in Guangzhou, offered a brief history of this development.[23] Before this 'cultural turn', mathematics education research was in a similar condition to the state of philosophy of mathematical practice at the moment of publication of *18 Unconventional Essays*, that is, mostly focused on the psychology of cognition in individuals. Now, following this 'cultural turn', researchers in mathematics education increasingly study the learning and teaching of mathematics as cultural activities. This is not quite the same as treating mathematical research as a cultural activity, so it remains to be seen whether notions drawn from this source could help philosophers of mathematical practice. Education is a natural context in which to deploy definitions of the sort we encountered in the anthropological literature, because such definitions often define culture in terms of non-biological reproduction, and even those anthropologists that do not mention reproduction in their definitions of culture regard reproduction as central to their enquiries. Philosophers of mathematical practice tend to focus more on the production of mathematical knowledge than on its reproduction.

In the presentations in the London series on mathematical cultures, those contributors who drew articulations of the notion of culture from mathematics education mostly reached for one of two theorists: Gert Hofstede and Alan Bishop.[24] Bishop gave the keynote address at the second conference in the London series. The remainder of this essay will sketch some of their ideas and offer some critical remarks.

4 Hofstede

Hofstede starts from the thought that teacher and student form a 'role-pair', and the effectiveness of this dyad can suffer if they come from different cultures that engender different expectations about what those roles require. He originally offered a four-dimensional frame for comparing national cultures (not necessarily

[23]https://youtu.be/umuKvJFR_7U see also François, Karen & Stathopoulou, Charoula (2012). 'In-Between Critical Mathematics Education and Ethnomathematics. A Philosophical Reflection and an Empirical Case of a Romany Students' group Mathematics Education.' *Journal for Critical Education Policy Studies, 10*(1), 234–247 ISSN 1740-2743.

[24]Paul Andrews made extensive reference to Hofstede's scheme in his international comparison of school mathematics teaching; Albrecht Heeffer used part of Bishop's framework in his presentation of the Abbaco mathematical culture. Slides and video recordings of their talks and Bishop's keynote address are available on the Mathematical Cultures project website https://sites.google.com/site/mathematicalcultures/.

in mathematics).[25] Since then, he has added two more dimensions[26] and extended the analysis from national cultures to organisational cultures.

The original four dimensions are:

Individualism/collectivism (the extent to which individuals act and think of themselves as members of in-groups larger than the immediate family unit)
Power distance (the extent to which the less powerful accept inequality)
Uncertainty-avoidance/-tolerance
Male/Female (the extent to which possible male roles overlap with possible female roles).

Hofstede presents these as culturally neutral categories. However, it is implausible that cultural theorists should somehow escape the scope of the founding premise of the cultural approach. Everything humans do is cultural, including cultural analyses. So, for example, Hofstede articulates the first item on this list (individualism/collectivism) in a way that treats the tight nuclear family as natural. In this picture, people who act on behalf of or as representatives of their immediate nuclear families do not thereby show themselves to be culturally collectivist (that is, in this model, acting for your sibling or parents counts as acting for yourself, but acting for your cousin or aunt does not). This is odd when we consider the much greater importance of the extended family over the nuclear in many, perhaps most cultures, and the relatively high incidence of parental mortality for most of human history. It rather looks like a feature of Hofstede's cultural origin has shaped his model.

Similarly, the last dimension of the four looks ripe for revision. It locates a culture on this framework's male-female scale by measuring the extent to which it is acceptable for men to undertake activities that are, within that culture, considered female. This builds in an assumption that either (a) women never do anything outside their cultural gender-roles, or (b) when they do, it is not interesting for social science. Hofstede argues that this assumption is in the data rather than in his approach (p. 308). Nevertheless, it is hard to see why one could not replace this dimension with a symmetrical notion such as gender-rigidity. That would capture the same data, without building sexist assumptions into the scientific apparatus.

---

[25] *International Journal of Intercultural Relations* Vol 10 pp. 301–320, 1986. In an earlier paper, Hofstede defines culture as "...the collective programming of the mind which distinguishes the members of one category of people from another." (p. 51). Hofstede, G. (1984). 'National cultures and corporate cultures'. In L.A. Samovar & R.E. Porter (Eds.), *Communication Between Cultures*. Belmont, CA: Wadsworth. Note that this definition is about categorising people rather than reproducing ideas (though the ideas are there, under the computer metaphor).

[26] He is not inconsistent in this; he prepared for this eventuality when writing the 4-dimensional model, "There is nothing magic about the number of four dimensions..." (p. 306).

Since then, Hofstede has added *Short-term/long-term* and *Indulgence/restraint*.

The fact that this is an open-ended list deprives it of some of its theoretical interest and power. Compare the achievement of psychologists in identifying *precisely* five dimensions of human personality. Or in philosophy, recall that when Kant offered 12 fundamental categories in four groups of three, he had a painstaking explanation why there are just that many in just that order.[27] In contrast, Hofstede could easily add a seventh, eighth and *n*th item as his curiosity roves, which leaves us wondering why we should classify societies in just these terms. We might find it more insightful to analyse a society in terms of a distinction between action and knowledge,[28] or according to the manifestation of *ressentiment,*[29] or rationalisation/disenchantment.[30] In such a case, Hofstede could easily protect his scheme from refutation just by adding an extra dimension—and so Popper's familiar point applies.

Nevertheless, this list of 'dimensions' is at least a list of questions that we can ask about mathematical cultures. For example, in 2012 at the first meeting in London, Slava Gerovitch described the Gelfand Seminar at Moscow University. Israel Gelfand made a point of scrambling the usual formalities and hierarchies. Invited Speakers could be interrupted or replaced on the spot and woolly explanations dismissed without mercy, regardless of the status of the speaker. Sessions always started late, but not always by the same delay—it might be minutes or hours—and they often went on until the building closed. The participants included distinguished professors and high-school students. Perhaps this meant that it had a low power-distance (since participants did not expect distinctions of rank to influence the discussion) or perhaps it had high power-distance, because a senior and rather powerful personality dominated. Gelfand usually ended up taking over the presentation. This example suggests that Hofstede has identified some useful questions, but perhaps these are not dimensions in any sense analogous to spatial dimensions or the dimensions of human personality.

5 Bishop

Alan Bishop set out his view in *Mathematical Enculturation: A Cultural Perspective on Mathematics Education* (Kluwer, 1991). Drawing on the work of the anthropologist Leslie A. White (pp. 60–62), he claims that three elements give

[27] *Critique of Pure Reason* (1787) A80/B106ff for the table of categories; A70/B95ff for the 'clue' found in logical theory; A95-130/B129-169 for his detailed account.

[28] See Arnold (1875) *Culture and Value* chapter iv 'Hebraism and Hellenism'.

[29] See Nietzsche, principally the *Genealogy of Morals* (1887).

[30] See Max Weber (1930) *The Protestant Ethic and the Spirit of Capitalism.*

cultures their distinctive characters: ideology, sentiment and sociology.[31] These terms are themselves vague and shifting, so the best procedure is to see what he does with them. Bishop quotes Kroeber and Kluckhohn on the centrality of values for understanding cultures (so he is with the idealists rather than the empiricists among social scientists). Under each of these three headings, he finds two values of mathematics. It is not clear from his text whether we should always expect a culture to have precisely two values under each of these headings. The values he finds in mathematics do not form dialectical poles (except possibly openness and mystery, which may oppose each other), nor do they seem to have a conceptual or genus/species relation with the three headings they stand under. This may mean that, as with Hofstede's scheme, we can add sub-categories if necessary (though one would expect to add new values under one of the existing three top-level categories). In short, Bishop has not done for the fundamental categories of mathematical value what Kant attempted to do for the basic categories of experience. Since Kant was never wholly satisfied with his own effort to found his categories, we might wonder how far to reproach Bishop for his lack of conceptual system. In any case, here is Bishop's scheme of values for mathematics:

Ideology: Rationalism & Objectism
Sentiment: Control & Progress
Sociology: Openness & Mystery

Taking Bishop's six characteristics of mathematics in turn:

Ideology: Rationalism within mathematics, for Bishop, seems to be the requirement of proof in the broad sense of mathematical reason-giving (so this can include the informal justification of procedures in mathematical traditions directed at problem-solving rather than theorem-proving). There is an intended contrast with cultures in which claims are established by appeal to tradition and/or authority. Under this heading, Bishop also discusses the role of mathematics within rationalism in a much larger sense, akin to rationalisation as described by Max Weber (that is, a deep historical process of systematisation, intellectualisation, standardisation and increasingly calculated efficiency in everything from the stacking of food on shelves to the doctrines and rituals of religion). This second thought bears on the meaning of mathematics for the rest of society. These two senses of rationalism come apart, because it is quite possible to be a mathematician and remain indifferent to the role

[31]This essay focuses on Bishop's third chapter, 'The Values of Mathematical Culture'. Heeffer's discussion (alluded to above) draws on other parts of Bishop's book. White's scheme included technological values as a separate category; Bishop does not. White's magnum opus is (1959) *The evolution of culture: the development of civilization to the fall of Rome*. New York: McGraw Hill.

of mathematics in Weberian rationalisation, or even, like Hardy, to make a point of doing mathematics that has no such role.[32]

Ideology: 'Objectism' is Bishop's word for mathematics' preference for fixed objects over processes and flows, and also over human relations. Bishop may have the history of the infinitesimal calculus in mind, which illustrates this in some readings (the early calculus referred to moving points and vanishing quantities, but these have been replaced by a standard continuum composed of stationary and ontologically stable points). It is hard to know how hard Bishop would wish to push this claim. There are mathematical models of processes, flows and human relations, and the fact that these require more advanced mathematics than counting discrete and stable objects or measuring definite fixed magnitudes may not be a matter of 'ideology' however this word is understood.

Also folded in 'objectism' is mathematics' tendency to reification (think of the way in which functions, for example, start out as relations between objects but become objects with properties, relations and mappings of their own).

Under 'objectism', Bishop includes a claim that mathematics engages more easily with the inanimate world, and most especially the world of manufactured inanimate objects. There are several points run together here: that mathematics deals most easily with manufactured objects is a different thought from its tendency to reification. Indeed, having distinguished these two ideas, we might relate them by suggesting that mathematics relates most easily of all to objects it has manufactured itself, that is, to reified mathematical objects. Mathematics, says Bishop, requires people to learn to treat abstract entities "*as if they were objects*" (p. 80, emphasis in original), which suggests an unargued philosophical conviction that abstracta are not, really, objects. Through all this, 'object' seems to mean 'material object'.

Bishop is of course not the first to notice this (nor does he claim to be). Here is C.S. Peirce:

> ... mathematical reasoning (which is the only deductive reasoning, if not absolutely, at least eminently) almost entirely turns on the consideration of abstractions as if they were objects.[33]

In reading this, we have to remember that Peirce had his own conception of deductive reasoning. Nevertheless, we have here an explanation-sketch for much that Bishop records under 'objectism'. If Peirce is right, then mathematical thinking is possible only on condition of treating abstractions as if they were objects, and

[32]Hardy, G.H. *A Mathematician's Apology.* (2004) [1940]. Cambridge: University Press. Note the date! Hardy published his thoughts on the uselessness of mathematics at the outbreak of war.

[33]Peirce, C.S. 'The logic of relatives' *The Monist* Vol. VII. January, 1897. No. 2. p. 192. I learned of this passage from Hacking, I. *Why is there philosophy of mathematics at all?* Cambridge: Cambridge University Press, 2014, p. 255.

explains why mathematics models some objects and phenomena more easily than it does others.

Sentiment: Control Here, Bishop talks more about techno-science than about pure mathematics, so it is not clear that this is a specifically mathematical feature. Mathematics may be useful for dominating and controlling but it is not obvious that this is something that mathematicians or mathematical cultures always value (recall the example of Hardy). There is a historical story to tell about European mathematics in this regard, from the conception of mathematics as the most abstract part of the study of nature (dominant in the eighteenth century) to the view of pure mathematics as art for art's sake that emerges at the end of the nineteenth century.[34] This discussion overlaps with some of the 'ideological' values gathered under rationalism in the form of Weberian rationalisation. Perhaps these are the same value, appearing first as ideology and now as sentiment. This is not the only overlap between Bishop's three main categories, and indeed, this is what one would expect. Members of religious communities, for example, experience their faith as doctrine, as sentiment and as a social order, and if a faith is coherent, one would expect to find the same values present in all three of its aspects. Similarly, in a mathematical culture in good order, we should expect to find the same values expressed in thought, in feeling and in social organisation—that is, as ideology, sentiment and sociology.

Sentiment: Progress Bishop's discussion here seems to mix progress in mathematical research with rising mastery in the student. Students can progress in their studies of anything, even in a subject that does not value innovation or progress in research. They can enjoy rising mastery in memorising received content, say in technical instruction or remembering a closed religious text. On the other hand, valuing progress in research is certainly part of contemporary research culture, but this is not distinctive of mathematics. Moreover, the cultures of mathematics in school classrooms seem to have little connection with progress in mathematical research (Snezana Lawrence made this claim about British school classrooms in her presentation to the first London conference on mathematical cultures[35]). Research and school-teaching have different interests, and these can set researchers and educationalists at odds (see Paul Ernest's useful table and analysis of these differences in the recent British context[36]).

Sociology: Openness Mathematics is open to anyone who can understand it. Mathematics is not tied to any religion or national culture, and for members of otherwise marginalised groups, this is can be a benefit. Authority in mathematics resides in

[34] See, for example, Amir Alexander's *Duel at Dawn*. Harvard University Press, 2010.

[35] https://youtu.be/FE6m-z61VTs, accessible from https://sites.google.com/site/mathematicalcultures/.

[36] *Why Learn Maths?* London: Institute of Education, 2000, pp. 6–7.

logical argument rather than in status, tradition or charisma, and these arguments are publicly available. In this section (pp. 75–77), Bishop gives mathematics a value system redolent of Enlightenment, especially the French Enlightenment (we already saw a little of that under Rationalism). Mathematics is democratic, open, subversive of arbitrary power, and so on. Of course, mathematicians and mathematical institutions do not always behave this way; results, methods and proofs may be hidden, autocratic professors or politically interested committees may kill lines of research. Bishop may reply that that these closed procedures can never prevail, other than in the short term. He ends this section by quoting Habermas, who urges that political power be rationalised (in Weber's sense) through dialogue. By this point, we have passed from resolutely value-free social science into a conception of mathematical culture as an educative example of the value of being "logical, precise, critical and argumentative" (p. 77). This is just what our earlier argument predicted about any analysis of mathematical cultures aimed at improving existing practice. Bishop does not entertain the thought that, for people living in harsh environments, group cohesion may be essential for survival, and in such circumstances being sociable, accommodating and respectful of established status may be more valuable than being "logical, precise, critical and argumentative". If Bishop is right that these values are intrinsic to mathematics, then in some circumstances people may have good reasons for rejecting mathematics, not on account of its technical content but rather because of the socially corrosive effect of the very values that Bishop extols.

Sociology: Mystery Bishop articulates this as the mysterious status of mathematics to the uninitiated, arising from the abstract nature of mathematical entities. It is not obvious that this esoteric status is something that mathematicians or mathematical cultures always value. Moreover, it cuts against the Enlightenment-spirit claims made under Rationalism and Openness (though this may be a dialectical opposition inherent in mathematical culture rather than a flaw in Bishop's scheme).

There are various sorts of mysticism internal to mathematics in addition to the esoteric status of mathematics. There are sophisticated versions of popular superstitions (lucky 7, unlucky 13, and so on). There are religious mysticisms, such as the Russian religious practice of name-worshipping, which may have played a role in some mathematical research.[37] In European intellectual culture, the most pervasive and influential form of mathematical mysticism originates with Pythagoras and Plato. Mathematical training, according to Plato, was a means (perhaps the principal means) by which the mind can prepare for communion with the unchanging intelligible forms that constitute the higher, nobler reality of which empirical reality is but a crude and shifting copy. Mathematics, in other words, can

[37] 'Mathematics and Mysticism, Name Worshipping, Then and Now' Jean-Michel Kantor *Theology and Science* Vol. 9, Issue 1, 2011.

function as a spiritual discipline like meditation, yoga or chanting.[38] However, in Plato's vision, mathematics can only be a bridge to the intelligible, because it is tethered to empirical experience.[39] Here we see an intellectual version of the esoteric tendencies in ancient Greek religion. This is quite different from what is called 'Platonism' in contemporary philosophy of mathematics. Contemporary Platonism may make a mystery of mathematical knowledge, but it does not value mystery.[40] There may be tendencies towards mysticism among some mathematicians, but it is not obvious that this says anything about mathematics.

Overall, Bishop's three-by-two matrix of mathematical values is supposed to express the universal values of mathematics, and one might wonder whether even Western mathematics of the last few centuries is quite so stable. Tracking the emergence of the distinction between pure and applied mathematics, as Ian Hacking does,[41] reveals that 'mathematics' has meant quite different things to different people, even in the relatively narrow pool of professional European mathematicians.

This thought is not new. One of the few early cultural theorists to write about mathematics was Oswald Spengler, who began his grand jeremiad on western civilisation with a chapter on mathematics:

> *There is not, and cannot be, number as such.* There are several number-worlds as there are several Cultures. We find an Indian, an Arabian, a Classical, a Western type of mathematical thought and, corresponding with each, a type of number—each type fundamentally peculiar and unique, an expression of a specific world-feeling, . . .[42]

In order to sustain this thesis in all its stark simplicity, Spengler immediately had to insist that apparent continuities between mathematical cultures are really ruptures. Historians of ancient mathematics will wonder how Indian, Arabian, Classical etc. mathematicians were able to learn from each other (compare the reported remarks of Netz and Azzouni, above). Historians of modern mathematics will marvel at Spengler's claim that western mathematics was (in 1918) in the same decedent, eventide condition as the rest of western culture, unable to produce anything really new, fated to do no more than extend and complete what it already had. (Spengler died in 1936; it seems unlikely that he spent his final years studying Van de Waerden's *Moderne Algebra*.) In general, Spengler's book is a model of the dangers of trumpeting the importance of particulars from a position of stratospheric

---

[38]Plato *Republic* 525 ff. For a useful discussion of this in a contemporary context, see Peter Huckstep 'Mathematics as a Vehicle for Mental Training' in *Why Learn Maths?* London: Institute of Education, 2000, pp. 88–91.

[39]*Republic* 511a.

[40]Though see Brian Rotman's claim that contemporary Platonism is a "theological obfuscation of number" (Ad Infinitum*: the ghost in Turing's machine.* Stanford: Stanford University Press, 1993, p. xii.

[41]Hacking, I. *Why is there philosophy of mathematics at all?* Cambridge: Cambridge University Press, 2014, chapter five.

[42]Oswald Spengler *The Decline of the West* Volume I: Form and Actuality. Tr. C. F. Atkinson. London: George Allen & Unwin Ltd., 1926, p. 59.

theoretical generality. This, together with its overall ideological orientation, advises us to treat it cautiously. However, his insistence on the differences between cultures (even in something apparently universal such as number) points to something latent in the definitions of culture that we started with. Cultural aspects of human life distinguish themselves from natural elements in that they vary in ways that have nothing to do with DNA. Whatever is universal is taken to be, in some sense, natural. For example, childbirth is natural and universal, but midwifery is various, changing and cultural. There is, therefore, a tension in Bishop's work. In urging a cultural approach to mathematics, he encourages attention to the variety of mathematical cultures, even as he sometimes writes as if there is really just one.

These, then, are the two articulations of the concept of culture that some participants appealed to at a meeting in London in 2012, and which seem to have some traction in mathematics education research. They are not tidy theories; Bishop's in particular seems to mix values held within mathematics with the value or image of mathematics in the host culture.

6 Final Thoughts

This rather roundabout discussion permits two conclusions.

First, there is no commonly agreed conception of culture that students of mathematical practices can take down from the shelf and use as an unproblematic tool. Mathematical practices certainly satisfy the definitions we considered: they are reproduced non-biologically, they are shared activities mediated by material artefacts, and they express norms and values. However, this raises questions about how these elements relate to each other that will affect the objects and the design of cultural studies. For example, we have to decide how much emphasis to put on the material traces of mathematics, and what inferences we think such traces can support.[43] There is also the problem, not discussed here, of the individuation of mathematical cultures (assuming we eschew both Bishop's mono-cultural view and Spengler's use of conventional Eurocentric divisions of human cultures into Indian, Asian, Classical, etc.). How, for example, would we answer this question: does theoretical computer science constitute a culture in an anthropologically robust sense?

Second, normativity is unavoidable. This is obvious in mathematics education research, aimed as it is at better learning and teaching. Even where we are not concerned with education, Netz's point applies. Philosophers are interested in mathematics as a successful practice, and historians are almost exclusively interested in successful episodes, individuals and institutions. Even if we ask why

[43]See, for example, Greiffenhagen, C. (2014), 'The materiality of mathematics: Presenting mathematics at the blackboard'. *The British Journal of Sociology*. The abstract begins "Sociology has been accused of neglecting the importance of material things in human life and the material aspects of social practices." and goes on to describe a 'material turn' in sociology.

some communities do not develop advanced mathematics, or why some institution did not deliver on its promise, we do so informed by norms of success and failure. This is hardly surprising. Compare studies of war, for example. These, inevitably, tend to focus on societies that are or were successful at it. This inevitable presence of normativity is not something to regret. It relieves philosophers of the duty of maintaining a fact-value distinction that is unstable, unachievable and possibly non-existent (depending on how hard one wishes to press the argument).

Appendix: Recent Conferences on Mathematical Cultures

24–26 October 2002. Perspectives on Mathematical Practices, Brussels, Belgium.

2006–2010. The PhiMSAMP network funded by the Deutsche Forschungsgemeinschaft with events and nodes in seven countries:

1. *PhiMSAMP-0*: Bonn, Germany, 8 May 2005.
2. GAP.6 Workshop: Towards a new epistemology of mathematics (= *PhiMSAMP-1*). Berlin, Germany, 14–16 September 2006.
3. *PMP 2007*. Perspectives on Mathematical Practices II, Brussels, 26–28 March 2007.
4. *PhiMSAMP-2*. Utrecht, The Netherlands, 19–21 October 2007.
5. *18th Novembertagung*.[44] Bonn, Germany, 1–4 November 2007.
6. *PhiMSAMP-3*. Vienna, Austria, 16–18 May 2008.
7. Foundations of the Formal Sciences VII. Brussels, Belgium, 21–24 October 2008.
8. *PhiMSAMP-4*. Brussels, Belgium, 24–25 October 2008.
9. *PhiMSAMP-5*. Hatfield, UK, 29–30 June 2009.
10. *Two Streams in the Philosophy of Mathematics: Rival Conceptions of Mathematical Proof*. Hatfield, UK, 1–3 July 2009.
11. PhiMSAMP-6. Utrecht, The Netherlands, 22–23 April 2010.

27–29 May 2010. Mathematics as Culture and Practice, Bielefeld, Germany.

9–11 December 2010. First International Meeting of the Association for the Philosophy of Mathematical Practice, Brussels, Belgium

2–3 December 2011. Mathematics as Culture and Practice II, Greifswald, Germany.

2010–2012. Symposia on mathematical practice and cognition at the conventions of the Society for the Study of Artificial Intelligence and Simulation of Behaviour.

1. Symposium on Mathematical Practice and Cognition, Leicester, UK, 29–30 March 2010.

[44] The *Novembertagung* series of meetings is intended for young historians and philosophers of mathematics.

2. Symposium on Mathematical Practice and Cognition II, Birmingham, UK, 2–4 July 2012.

2012–2014. Mathematical Cultures: workshop series funded by the UK Arts and Humanities Research Council and the London Mathematical Society:

1. *Mathematical Cultures 1: Contemporary mathematical cultures*, London, UK, 10–12 September 2012.
2. *Mathematical Cultures 2: Values in mathematics*, London, UK, 17–19 September 2013.
3. *Mathematical Cultures 3: Mathematics in public culture*, London, UK, 10–12 April 2014.

9–12 November 2012. Cultures of Mathematics and Logic, Guangzhou, China (the origin of this book).

20–23 September 2013. Foundations of the Formal Sciences VIII: History and Philosophy of Infinity, Cambridge, UK.

3–4 October 2013. Second International Meeting of the Association for the Philosophy of Mathematical Practice Urbana-Champaign, USA.

22–25 March 2015. Cultures of Mathematics IV New Delhi, India.

2–4 November 2015. Third International Meeting of the Association for the Philosophy of Mathematical Practice, Paris, France.

Most of these meetings have associated books of proceedings.

References

Adam, J. (Ed.). (1902). *The Republic of Plato. Edited, with critical notes, commentary, and appendices* (2 Vols.). Cambridge: Cambridge University Press.

Alexander, A. (2010). *Duel at dawn*. Cambridge, MA: Harvard University Press.

Alrø, H., Ravn, O., & Valero, P. (Eds.). (2010). *Critical mathematics education: past, present and future*. Rotterdam: Sense Publishers.

Arnold, M. (1978). *Culture and anarchy*. Cambridge: Cambridge Univ. Press. Reprint of the second (1875) edition.

Azzouni, J. (2006). How and why mathematics is unique as a social practice. In R. Hersh (Ed.), *18 unconventional essays on the nature of mathematics*. New York, NY: Springer.

Banks, J. A., & McGee Banks, C. A. (1989). *Multicultural education*. Needham Heights, MA: Allyn & Bacon.

Bishop, A. (1991). *Mathematical enculturation: a cultural perspective on mathematics education*. Dordrecht: Kluwer.

Ernest, P. (2000). Why teach mathematics? In J. White & S. Bramall (Eds.), *Why learn maths?* London: London University Institute of Education.

François, K., & Stathopoulou, C. (2012). In-between critical mathematics education and ethnomathematics. a philosophical reflection and an empirical case of a Romany students' group mathematics education. *Journal for Critical Education Policy Studies, 10*(1), 234–247.

Greiffenhagen, C. (2014). The materiality of mathematics: presenting mathematics at the blackboard. *The British Journal of Sociology, 65*, 502–528.

Hacking, I. (2014). *Why is there philosophy of mathematics at all?* Cambridge: Cambridge Univ. Press.
Hardy, G. H. (2004). *A mathematician's apology*. Cambridge: Cambridge Univ. Press [Original edition 1940].
Hegel, G. W. F. (1821). *Elements of the philosophy of right*. (Original German title: *Grundlinien der Philosophie des Rechts*).
Hersh, R. (2006). *18 unconventional essays on the nature of mathematics*. New York: Springer.
Hofstede, G. (1984). National cultures and corporate cultures. In L. A. Samovar & R. E. Porter (Eds.), *Communication between cultures*. Belmont, CA: Wadsworth.
Hofstede, G. (1986). Cultural differences in teaching and learning. *International Journal of Intercultural Relations, 10*(3), 301–320.
Huckstep, P. (2000). Mathematics as a vehicle for mental training. In S. Bramall & J. White (Eds.), *Why learn maths?* (pp. 88–91). London: Institute of Education.
Kant, I. (1787). *Critique of pure reason*. (Original German title: *Kritik der reinen Vernunft*) first published in 1781, second edition 1787. Riga: Johann Friedrich Hartknoch.
Kantor, J.-M. (2011). Mathematics and mysticism, name worshipping, then and now. *Theology and Science, 9*, 1.
Kroeber, A. L., & Kluckhohn, C. (1952). *Culture: a critical review of concepts and definitions*. Cambridge, MA: Peabody Museum of American Archeology and Ethnology Harvard University. Papers 47.
Larvor, B. (2010). Review of the philosophy of mathematical practice. In P. Mancosu (Ed.), *Philosophia mathematica* (Vol. 18, pp. 350–360). Oxford: Oxford University Press.
Larvor, B. (2012). The mathematical cultures network project. *Journal of Humanistic Mathematics, 2*(2), 157–160.
Lockhart, P. (2009). *Mathematician's lament*. New York, NY: Bellevue Literary Press.
Mancosu, P. (Ed.). (2008). *The philosophy of mathematical practice*. Oxford: Oxford University Press.
Manders, K. (2008). The Euclidean diagram. In P. Mancosu (Ed.), *The philosophy of mathematical practice* (pp. 80–133). Oxford: Oxford University Press.
Netz, R. (1999). *The shaping of deduction in Greek mathematics*. Cambridge: Cambridge Univ. Press.
Nietzsche, F. (1887). *On the genealogy of morality: a polemic*. (Original German title: *Zur Genealogie der Moral: Eine Streitschrift*). Leipzig: C.G. Naumann.
Parsons, T. (1949). *Essays in sociological theory*. Glencoe, IL: Free Press.
Peirce, C. S. (1897). The logic of relatives. *The Monist, 7*(2), 161–217.
Rav, Y. (1999). Why do we prove theorems? *Philosophia Mathematica, 7*(1), 5–41.
Brian, R. (1993). *Ad infinitum: the ghost in Turing's machine*. Stanford: Stanford University Press.
Rosental, C. (2008). *Weaving self-evidence: a sociology of logic*. Princeton, NJ: Princeton University Press. Trans. by Catherine Porter.
Schiller, F. (1794). *Letters on the aesthetic education of man*. (Original German title: *Über die ästhetische Erziehung des Menschen*).
Skovsmose, O. (1994). *Towards a philosophy of critical mathematics education*. Dordrecht: Kluwer.
Spengler, O. (1926). *The decline of the west, Vol. 1: form and actuality*. London: G. Allen. Trans. C. F. Atkinson.
Useem, J., & Useem, R. (1963). Men in the middle of the third culture: the roles of American and non-western people in cross-cultural administration. *Human Organizations, 22*(3).
Weber, M. (1930). *The protestant ethic and the spirit of capitalism*. (Original German title: *Die protestantische Ethik und der Geist des Kapitalismus*). London: Allen & Unwin [Translated by Talcott Parsons].
White, J., & Bramall, S. (Eds.). (2000). *Why learn maths?* London: London University Institute of Education.

White, L. A. (1959). *The evolution of culture: the development of civilization to the fall of Rome*. New York: McGraw Hill.
Wilder, R. L. (1965). *Introduction to the foundations of mathematics* (2nd ed.). New York: John Wiley (First published 1952).
Wilder, R. L. (1981). *Mathematics as a cultural system*. Oxford: Pergamon.

Philosophy or Not? The Study of Cultures and Practices of Mathematics

Benedikt Löwe

Abstract
The most commonly accepted name of our research field is *Philosophy of Mathematical Practice*, giving philosophy a prioritized role among the many disciplines involved in the field. We explore the interplay between philosophy and other disciplines and its effect on the further development of our field.

1 Introduction

The conference *Cultures of Mathematics and Logic* in Guangzhou brought together philosophers, sociologists, historians, cognitive scientists, and researchers in mathematics education; it was one event among many in the past decade that studied

This paper was finished whilst the author was a Simons Visiting Fellow of the Programme *Mathematical, Foundational and Computational aspects of the Higher Infinite* at the Isaac Newton Institute for Mathematical Sciences (supported by the Simons Foundation and EPSCR grant EP/K032208/1). The author would like to thank Brendan Larvor, Thomas Müller, and Bart Van Kerkhove for comments.

B. Löwe (✉)
Institute for Logic, Language and Computation, Universiteit van Amsterdam, Postbus 94242, 1090 GE Amsterdam, The Netherlands

Fachbereich Mathematik, Universität Hamburg, Bundesstraße 55, 20146 Hamburg, Germany

Corpus Christi College, University of Cambridge, Cambridge CB2 1 RH, England

Isaac Newton Institute for Mathematical Sciences, 20 Clarkson Road, Cambridge CB3 0EH, England
e-mail: b.loewe@uva.nl

S. Ju et al. (eds.), *Cultures of Mathematics and Logic*, Trends in the History of Science, DOI 10.1007/978-3-319-31502-7_2

cultures and practices of mathematics. The appendix of (Larvor, 2016) in this volume lists these events.

A look at this list reveals that many of them have titles that include the word "philosophy" or philosophical terms such as "epistemology". The community that meets at these events is closely linked to the *Association for the Philosophy of Mathematical Practice* (APMP) which serves as the institutional backbone of the field. Many of the researchers in our field refer to it as *Philosophy of Mathematical Practice* and self-identify as philosophers. A conference participant who is not a philosopher might ask: why does philosophy play such an important role in the field? In contrast, are those events that do not grant philosophy this priority even events in the same field?

In this paper, we shall explore the tension between the central role of philosophy in our community and the fact that philosophy is only one among many fields interested in mathematical cultures and practices. We start in Sect. 2 by exploring the notions of "research field", "community", and "discipline" and argue that, entirely independently of which name we choose for it, we can talk of "our community", even though its members belong to different disciplines and may have different motivations. In Sect. 3, we turn to the name of our community and propose that the occurrence of the term "philosophy" in "philosophy of mathematical practice" indicates a specific goal of research rather than a claim that the study of mathematical cultures and practices is intrinsically philosophical in nature and explain that the philosophically motivated research on mathematical practices and cultures has been a driving force for our community. In the following two sections, we then discuss the practical consequences of this contingent fact: in Sect. 4, we discuss a number of important distinctions for empirical approaches to philosophy, in particular one due to Prinz; in Sect. 5, we then discuss how the overall research situation in the broader field of studying mathematical practices and cultures has an effect on which of Prinz's approaches we can choose. We conclude in Sect. 6 with a discussion about the consequences of the analysis of our field given in this paper.

2 Our Community

Several factors, both intrinsic and extrinsic to the academic and scientific content, are relevant for determining whether a field of research can be classified as a *research community* or even a *discipline*.[1] Among the intrinsic factors are coherence in the subject matter of research, the motivation of the research, and the methods of doing the research; the extrinsic factors include questions such as whether there are

[1]In the following, we shall not define either of the three terms "field of research", "research community", or "discipline" precisely. We shall be using "field of research" as the most generic term covering any collection of researchers or research results; the other two terms, "research community" and "discipline" are more specific, but we remain agnostic about their precise definition, as the discussion in this section exhibits.

research institutes, departments, research programmes, conference series, journals, undergraduate and postgraduate teaching programmes for the field. Sometimes, the status of a field as a discipline is being questioned on some grounds, e.g., that the field does not have a homogeneous research method or its own publication culture. In many cases, when publicly raised, these doubts come with a strategic agenda, either as an attempt to discredit a field or as a rallying cry from inside the field.

In some relevant debates, both the disciplinary nature of a research field and some of the intrinsic factors for disciplinarity become crucial: for instance, in a paper that became the starting point for a lively debate, Stokhof and van Lambalgen (2011) raised the question whether "modern linguistics [is] [...] an example of a 'failed discipline' " (p. 3). Answering such a question presupposes that modern linguistics is a discipline that shares a common list of goals in comparison to which the achievements of the field can be seen as "failure". Maddirala (2014) discusses the issue raised by Stokhof and van Lambalgen for the more narrow field of formal semantics by doing an interview study with formal semanticists; his study shows that discrepancies between the goals and intentions attributed to a field from the outside, those professed by members of the community in public, and those implicitly observable from work of the members of the field make it difficult to determine the precise meaning of questions like the one by van Lambalgen and Stokhof.

2.1 Subject Matter

Taking philosophical issues seriously, uniformity of the subject matter of the field is difficult to achieve since agreement about what the subject matter is presupposes agreement about major questions about the ontology of the field; e.g., saying that mathematics is the study of properties of mathematical objects is taking a strong philosophical position about the ontology of mathematics. Even in established subfields of philosophy such as philosophy of language and philosophy of mind, it would be difficult to get the practitioners of the field to agree on the ontology of their area's subject matter. The only solution to this problem is to suspend the philosophical commitment to scrutinize fundamental terms and leave the subject matter deliberately philosophically underdetermined by calling it, e.g., "language" or "the mind" without giving a definition.

Disciplinary borderlines do not respect the boundaries of metaphysical schools: mathematicians who believe that mathematical objects exist independently from human activity and those who believe that mathematics is just a formal game are equally classified as "mathematicians" and neither of them would refuse to call the other a mathematician. So, when we talk about "subject matter" in the context of the definition of disciplines, we are talking about terms that are philosophically underdetermined.

In our case, we can and do agree that the subject matter is "mathematical practices and cultures" or "mathematical activity as practised by human beings". As soon as we apply philosophical scrutiny to this underdetermined description, the consensus of what the subject matter of our field is will break down, but our field does not differ in this respect from established disciplines such as mathematics, physics, or others.

2.2 Research Methods

It is primarily the uniformity of the research method or the collection of research methods that determine a discipline. Etymologically, the word discipline suggests the body of knowledge or techniques that are taught to the *discipuli* in a rigourous teaching environment:

> [*disciplina*] désigne [...] ce qui fait d'objet de l enseignement, la MATIÈRE ENSEIGNÉE, μάθησις, et spécialement la matière d'un enseignement régulier, systématique, autrement dit une science, ἐπιστήμη. (Marrou, 1934, p. 6)
> La notion d'enseignement devient très lontaine; elle finit par s'effacer et *disciplina* signifie règle imposée [et] BON ORDRE. (Marrou, 1934, p. 11)

By definition, interdisciplinary endeavours do not have this uniformity of research methods. Our field, driven by the realisation that one discipline alone cannot give a multidimensional picture of all facets of mathematical practice, is a prototype of such an interdisciplinary endeavour. As a consequence, we cannot and should not expect that the field is defined by methodological coherence. Assuming the standard usage of the word "discipline", we observe that our field cannot be a discipline.

2.3 Motivation

Even established disciplines do not have a commonly shared motivation. Taking mathematics as an example, representatives of different subfields of mathematics will strongly disagree on what the motivation for their work is. If you organise a panel discussion on the topic of 'What motivates mathematical research?' with a set theorist, a differential geometer, a mathematical physicist, and an actuary, you can expect heated debate and very little agreement. The dispute may even reach the point where some panelists start to doubt whether mathematics is a single discipline, given their fundamental dissensus. Yet, these emotional reactions do not become a serious argument that mathematics is not a discipline; instead, we have to revise our view that a uniform motivation is important for the disciplinary character of a field.

Similarly, we shall observe that members of our field strongly diverge on what their motivations for their research are, even to the point where some feel that the disagreement is so fundamental that we are not dealing with one community, but several. The discussion of our differing motivations for being interested in "mathematical practices and cultures" is the main focus of Sect.3, so we defer this discussion to later.

2.4 Extrinsic Factors

Many of the mentioned extrinsic factors (existence of research institutes, departments, research programmes, conference series, journals, undergraduate and postgraduate teaching programmes) are difficult to separate from the institutional set-up of our academic world; therefore, there is a strong correlation between extrinsic factors and whether a field is considered a traditional discipline or not. Since the vast majority of the universities, research institutes and funding agencies is organised along the lines of the established disciplinary boundaries, there is something that we could dub the *disciplinary barrier*: institutions make decisions whether to grant a degree, whether to accept a postgraduate student, whether to hire a postdoc or a new faculty member, whether to give tenure or to promote, whether to fund a research project, whether to award a prize or elect to a position of recognition mostly according to the standards of the relevant established discipline. This means that the participants of any interdisciplinary research endeavour will need to play according the disciplinary rules in order to maintain the chances of a research career. This makes it very difficult and—if the discussed field is interdisciplinary by design—possibly undesirable to establish some of the mentioned extrinsic factors that create community cohesion beyond or between the disciplines.

For instance, typically (but not always) a research journal requires submitted papers to use a particular research methodology. Of course, there are generalist journals (such as *Science* and *Nature*) or genuinely interdisciplinary journals (such as the newly founded journal *Computability*), but there are practical reasons why such examples are few and far between. Papers that straddle methodological boundaries are difficult to judge: If a paper uses methods from fields X and Y, does it have to meet the standards of publication in both X and Y, or do we think that the joining of techniques creates intrinsic added value going beyond the sum of its parts? Do we need referees from both X and Y? If they disagree, how do we aggregate their judgments and how do we deal with conflicting judgment styles between the disciplines?

Similar practical issues apply to other extrinsic factors such as postgraduate programmes (does a thesis merit an interdisciplinary degree in X and Y if it is not sufficient for a degree in either X or Y?) or departments and research institutes (do we want to hire a person who would not be strong enough to be hired in a Department of X or in a Department of Y?), and research fields that are interdisciplinary hit what we called the disciplinary barrier.

On the other hand, our field exhibits most, if not all extrinsic indicators for being a community that are just below the disciplinary barrier: conferences are being organised with proceedings volumes that are published as books or special issues of leading research journals, conference series and research networks, often with funding from standard funding agencies, are being established, and, not least, the members of our field decided to form an association, the *Association for the*

Philosophy of Mathematical Practice.[2] Those of us who have been to several of the conferences and workshops of the field realize that people they met at one of the conferences also come to others, allowing discussions to extend beyond the confines of a single conference. As usual in interdisciplinary fields, every participant thinks that they are on the fringe of the subject and that everyone else has more sense of belonging than they do; but meeting the same people in different configurations at various conferences gives a clear sense of cohesion.

We conclude that we are allowed to talk about a *community*: a field that coheres enough that community members collaborate, publish jointly, regularly meet at conferences and workshops, set up networks, but still remain part of their respective disciplines each with its own research and publication expectations. It is this community that we wish to understand better in the remaining sections of this paper.

3 Philosophy

In Sect. 2, we have seen that our field is not a *discipline* in the traditional sense, but exhibits many of the extrinsic indicators for being a *research community*. So far, we have been using "our field" to refer to it, deliberately avoiding to give it a name. In this section, we shall now move to the thorny issue of naming our community.

One of the important components of our argument that our field constitutes a community was the fact that the *Association for the Philosophy of Mathematical Practice* forms an institutional backbone for our field. That in turn could be used as an argument that "Philosophy of Mathematical Practice" is the official and accepted term for the field. Does that imply that our field is a subfield of philosophy? Many if not most of the members of our community are philosophers or are at least philosophically interested. This interesting contingent fact about our community asks for an explanation: why is it that philosophers are so interested in studying mathematical cultures and practice?

For many of the philosophers in our community, the reasons for becoming interested in the empirical study of mathematical practice were negative: the foundational debates in the philosophy of mathematics in the early twentieth century had resulted in a foundationalist school of thought dealing with a highly idealised version of what mathematics is, embedded in the formalism of first-order logic (or sometimes, second-order logic). Crucial notions of mathematical epistemology such as the notion of "proof" occurred in their sanitized form of formal derivations in a particular formal framework. The role of mathematical logic for the understanding of what mathematics can and cannot do should not be underestimated; but on the other hand, philosophers in our community felt that it was equally obvious that the amount of idealisation involved in the transformation from the complex human activity we call mathematics to the crisp and clean formal world of formal

[2]Once more, we refer to the appendix of (Larvor, 2016) for a list of these events and activities.

derivations hides many important and interesting features of what it is to do mathematics:

> [From the early 20th century onwards,] the foundations of mathematics became a central research interest. This resulted in a widely accepted notion of *formal derivation* as the explication of mathematical proof. [...] In mathematical practice, proofs are written down in a more condensed, semi-formal style. [...] The traditional view would consider these proofs *enthymematic*, leaving out technical detail for purely pragmatic reasons. [...]
> A closer look at mathematical practice leads to two important observations. First, the completion of enthymematic, semi-formal proofs to formal derivations almost never happens and hardly plays any rôle in the justification that mathematicians give for their theorems; second, also the production of semi-formal proofs [...] is only the final step of the mathematical research process. This final step, while important for the documentation of results and crucial for the careers of researchers, is not necessary for the acceptance of a proof by the mathematical community. For this, different forms of proof are much more relevant: informal sketches on the blackboard, or scribblings and drawing on napkins [...] Shouldn't these forms of proof replace the unrealistic notion of formal derivation in our epistemology of mathematics? (Buldt et al., 2008, pp. 310–311)

This attitude resulted in a number of publications whose very titles suggest that the traditional philosophical account does not deal with the core of what mathematics is about, e.g., *What is mathematics, really?* (Hersh, 1997) or *Towards a philosophy of real mathematics* (Corfield, 2003). The term "mathematical practices" was used by the Brussels philosophers of mathematics Jean Paul Van Bendegem and Bart Van Kerkhove when they started their series of meetings in 2002,[3] and the term "philosophy of mathematical practice" was chosen by Paolo Mancosu (2008b) as the title of a book containing papers by some of the protagonists of our community. The publication of (Mancosu, 2008b) has to be seen in the context of the formation of the *Association for the Philosophy of Mathematical Practice*, founded in the year 2009 with Mancosu as one of the nine founding members (three of the nine founders contributed to Mancosu, 2008b; three more are cited in Mancosu, 2008a). The Brussels conferences *Perspectives on Mathematical Practices* (October 2002 and March 2007) and their proceedings volumes (Van Kerkhove, 2008; Van Kerkhove et al., 2010; Van Kerkhove and Van Bendegem, 2007) were important contributing factors in the pre-history of the *Association for the Philosophy of Mathematical Practice*.

In his introduction, Mancosu (2008a) contrasts the philosophy represented in his volume with traditional philosophy of mathematics, giving an implicit definition of "philosophy of mathematical practice" in terms of this contrast:

> The contributions presented in this book are [...] joined by the shared belief that attention to mathematical practice is a necessary condition for a renewal of the philosophy of mathematics. (Mancosu, 2008a, p. 2)

[3]However, their acronym PMP stands for "Perspectives on Mathematical Practices" rather than "Philosophy of Mathematical Practice".

He then discusses the "foundational tradition" and the "maverick tradition" representing two movements within philosophy of mathematical practice. The term "maverick tradition" goes back to Aspray and Kitcher[4] and Mancosu characterizes it by

> a. anti-foundationalism, i.e., there is no certain foundation for mathematics; mathematics is a fallible activity; b. anti-logicism, i.e., mathematical logic cannot provide the tools for an adequate analysis of mathematics and its development; c. attention to mathematical practice: only detailed analysis of large and significant parts of mathematical practice can provide a philosophy of mathematics worth its name. (Mancosu, 2008a, p. 5)

One of the mentioned proponents of the maverick tradition, Jean Paul Van Bendegem (2014), also one of the founders of the Association, gives another list of eight approaches of the philosophy of mathematical practice:

> (a) the Lakatosian approach, also called the 'maverick' tradition; (b) the descriptive analytical naturalizing approach; (c) the normative analytical naturalizing approach; (d) the sociology of mathematics approach; (e) the mathematics educationalists approach; (f) the ethnomathematical approach; (g) the evolutionary biology of mathematics; and (h) the cognitive psychology of mathematics. (Van Bendegem, 2014, p. 221)

A third founder and the first president of the Association, José Ferreirós, drew a picture of the simultaneous diversity and unity of the field during a talk given on 11 September 2014 in Pont-á-Mousson:

> [Philosophy of Mathematical Practice] has different branches [that share] some very basic tenets [...] (the need to complement philosophical analyses with new features emerging from attention to concrete cases, the role for history and present-day studies, the emphasis on methodological differences between areas of math[ematics], the openness to interdisciplinary considerations, etc.) but then there are quite different ways of articulating [these tenets]. The main idea [is] [...] that philosophers have moved away from a static, monolithic, too idealised and simplified version of what mathematics is, towards what may be called 'real math[ematics]' in the sense of different aspects of the work and activities of mathematicians. In this [move] [...], some people have remained more cautious, while others have been more radical in taking into consideration the actual practice of doing math[ematics] by agents [...] or the social network of mathematicians. (Ferreirós, 2014, § 3)

[4] Kitcher and Aspray (1988, p. 17) describe the maverick tradition in terms very similar to the other characterisations of our community in this section: "[I]t is pertinent to ask whether there are [...] tasks for the philosophy of mathematics [...] that arise either from the current practice of mathematics or from the history of the subject. A small number of philosophers [...] believe that the answer is yes. Despite large disagreements among the members of this group, proponents of the minority tradition share the view that philosophy of mathematics ought to concern itself with the kinds of issues that occupy those who study other brnaches of human knowledge [...]: How does mathematical knowledge grow? What is mathematical progress? What makes some mathematical ideas [...] better than others? What is mathematical explanation?"

In the first chapter of his book *Mathematical Knowledge and the Interplay of Practices*, Ferreirós (2015) provides a manifesto of what could be called "epistemology of mathematical practice" and describes it as an interdisciplinary endeavour using tools from a plethora of disciplines and applying them to traditional questions from philosophy:

> During the twentieth century, we have seen several different broad currents in this field, which [...] can be reduced to three main types: *foundational* approaches [...], *analytic* approaches [...], and the so-called "*maverick*" approaches [...], which have typically been anti-foundational and focused on history, methodology, and patterns of change. [...] It seems to be the case that a new generation of philosophers of mathematics has arisen whose work is superseding those distinctions. [...] These philosophers engage in an analysis of mathematical practices that incorporates key concerns of the "mavericks", without adopting their anti-foundational, anti-logical orientation. [...] Notice that the new orientation in the philosophy of mathematics is highly interdisciplinary. Some authors emphasize knowledge of mathematics itself and logic [...]; some others stress the role of cognitive science [...] or sociological approaches [...]; and the list goes on, with mathematics education, anthropology, biology, etcetera. (Ferreirós, 2015, pp. 1–2)

The quotes by Mancosu, Van Bendegem, and Ferreirós emphasis that the philosophy of mathematical practice is not dealing with entirely new philosophical questions, but is a particular approach to philosophy of mathematics. In this context, it is interesting to note at the inaugural conference of the *Association for the Philosophy of Mathematical Practice* in Brussels there was a critical discussion of the term "philosophy of mathematical practice".[5] Its syntactic form "philosophy of *X*" suggests that there is an object "mathematical practice" whose philosophy it is studying. In particular, the name suggests that the field is distinct from "philosophy of mathematics": whereas the latter studies mathematics with philosophical means, the name "philosophy of mathematical practice" could suggest that the former studies the practice of mathematics, rather than mathematics itself.[6] This view was in general rejected by the participants of the inaugural conference; instead, the consensus was that philosophy of mathematical practice is an approach (or a collection of approaches) to philosophy of mathematics and this view is reflected in the definition of the purview of the *Association for the Philosophy of Mathematical Practice* on its webpage:

> Over the last few years approaches to the philosophy of mathematics that focus on mathematical practice have been thriving. Such approaches include the study of a wide variety of issues concerned with the way mathematics is done, evaluated, and applied, and in addition, or in connection therewith, with historical episodes or traditions, applications,

[5]The following argument was mentioned during a round table discussion during the inaugural conference on 10 December 2010.

[6]Other terms than "philosophy of mathematical practice" have been used that avoid this misinterpretation, among them "empirical philosophy of mathematics" (Löwe et al., 2010), "practice-based philosophy of mathematics" (Dutilh Novaes, 2012), "(socio-)empirically informed philosophy of mathematics" (Müller-Hill, 2009, 2011), or "philosophy of real mathematics" (Corfield, 2003).

> educational problems, cognitive questions, etc. We suggest using the label "philosophy of mathematical practice" as a general term for this gamut of approaches, open to interdisciplinary work.

But are all of the items listed as part of this "gamut of approaches" really approaches to philosophy of mathematics? In Van Bendegem's list of approaches for philosophy of mathematical practice, five of the eight items are

> not, strictly speaking, approaches 'in the philosophy of mathematical practice' [...] They are, rather, five *non*-philosophical perspectives on mathematical practice that are *used* by philosophers of mathematical practice or, more prudently, on which *some* philosophers of mathematical practice *can find relevant* to rely. (Jullien and Soler, 2014, p. 228; emphasis in the original)

In Sect. 2.1, we claimed that the subject matter of our field is "mathematical practices and cultures"; we now observed that there is consensus that philosophy of mathematical practice is not to be considered the philosophy of a separate subject matter, but rather as a particular approach in philosophy of mathematics. How do we reconcile these two contradictory observations?

The discipline of mathematics, seen as a human activity with its cultural particularities and achievements as well as various practices, is a "subject matter" in the philosophically underdetermined sense of Sect. 2.1 and well worthy of study; there are researchers from many disciplines studying this subject matter from various angles. For lack of a better term, one could call this research field *the study of mathematical cultures and practices*. Different researchers in this field of research have different motivations for studying mathematical cultures and practices: a researcher in mathematics education might be motivated by educational questions or even questions about educational policy; a cognitive scientist would be curious about understanding the cognitive processes in doing mathematics in contrast to other cognitive processes; an anthropologist could be fascinated by the difference between mathematical cultural practices and those of other, closely related, yet different disciplines; and, of course, the philosopher would be driven by traditional philosophical questions about epistemology and ontology of mathematics. Every researcher in the field must decide on the basis of the motivating questions which occurring phenomena are sufficiently relevant for their work.[7]

The two mentioned seemingly contradictory observations can be reconciled by understanding that they are two descriptions of extensionally very similar research communities: from the point of view of the broader interdisciplinary community that we dubbed *the study of mathematical cultures and practices*, our community is the sub-community motivated by philosophical questions; from the point of view of the larger philosophy of mathematics community internal to philosophy, our community is the sub-community that embraces a particular approach, basing

[7]Cf. (Löwe, 2014) for an example of a meta-argument for the philosophical relevance of a particular aspect of the practice of mathematics (viz. the use or the rejection of computer technology).

philosophical claims on actual mathematical practice rather than idealisations. As discussed in Sect. 2.3, the fact that the diverging motivations sometimes create feelings of division should not disturb us too much, as this happens in established disciplines as well.[8]

4 Three Relevant Distinctions for Empirical Philosophy

Jesse Prinz (2008) introduces a distinction between two approaches of doing philosophy on the basis of empirical data. He calls these approaches *Experimental Philosophy* and *Empirical Philosophy*:

> Some philosophers make use of empirical results that have been acquired by professional scientists. ... These results are used to support or refute philosophical theories. We shall call this approach 'empirical philosophy'. Other philosophers also conduct their own psychological experiments, an approach known as 'experimental philosophy'. (Prinz, 2008, p. 196)

His distinction is largely based on whether the empirical work is done by the philosophers themselves or rather by other scientists ("mining the data" vs. "collecting the data"):

> Empirical philosophy works by citation. Philosophers cite relevant empirical research and use it to argue for philosophical conclusions. (Prinz, 2008, p. 200)

Prinz acknowledges that the distinction is a contingent sociological fact about philosophy,[9] but argues that the two types of philosophers correspond to a natural division of types of philosophical questions:

> I am not trying to suggest that experimental and empirical philosophy *must* differ along the lines I suggest; only that they often do, and that there are reasons for these differences. (Prinz, 2008, p. 197)

We believe that the choice of terms *empirical philosophy* and *experimental philosophy* is infelicitous. There are at least three relevant distinctions with respect to the use of empirical data that should be considered here: there is a fundamental methodological question of whether philosophy should take empirical data about human activities and cognitive states into account; assuming that empirical data

[8]It is interesting to note the programmatic statements cited in this section implicitly or explicitly acknowledge these feelings of division: e.g., Kitcher and Aspray (1988) mention "large disagreements among the members of this group" and Ferreirós (2014) emphasizes that "there are quite different ways of articulating [these tenets]".

[9]Cf. also (Prinz, 2008, p. 200): "This distinction between experimental and empirical philosophy is very rough. ... [T]here are counter-examples in the literature, and the distinction is likely to blur even more in the years to come."

are taken into account, then there is a second methodological question of which empirical techniques are being used to collect this empirical data; and finally, there is the Prinzian question of whether the philosopher who asks the question is doing this empirical work himself or herself or "works by citation".

4.1 First Philosophy vs. Second Philosophy

Whether empirical data should be taken into account for philosophical arguments is the crucial question for the debate not only between experimental philosophers and their critics, but also between naturalistically-minded philosophers in general (not all of whom subscribe to the paradigm of experimental philosophy[10]) and traditionally-minded philosophers. In the meta-discussion about experimental philosophy, the latter position is called *armchair philosophy*. The main argument of armchair philosophers against the use of empirical data is that it is not really dealing with the philosophical concepts themselves, but rather with general, possibly extra-philosophical usage of the terms corresponding to the concept:

> [Since philosophers generally assume] competence of the speaker, absence of performance errors, and basis in semantic rather than pragmatic considerations [...], intuition statements cannot be interpreted as straightforward predictions, and therefore cannot, for reasons of principle, be tested through the methods of non-participatory social science, without taking a stance on the concepts involved and engaging in dialogue. For example, when philosophers claim that according to our intuitions, Gettier cases are not knowledge, they are not presenting a hypothesis about gut reactions to counterfactual scenarios but, more narrowly, staking a claim of how competent and careful users of the ordinary concept of knowledge would pre-theoretically classify the case in suitable conditions. The claim, then, is not about what I will call surface intuitions but about robust intuitions, which are bound to remain out of reach of the Survey Model of experimentalists. (Kauppinen, 2007, p. 97)

Maddy (2007) introduces the term "second philosophy" for her position that she called "naturalism" in earlier work (Maddy, 1997), as she feels that the term 'naturalism' "has come to mark little more than a vague science-friendliness (p. 1)"[11]:

> [The] Second Philosopher is equally at home in anthropology, astronomy, biology, botany, chemistry, linguistics, neuroscience, physics, physiology, psychology, sociology, ... and even mathematics, once she realizes how central it is to her ongoing effort to understand the world. [...] She simply begins from commonsense perception and proceeds from there to systematic observation, active experimentation, theory formation and testing, working all the while to assess, correct, and improve her methods as she goes. (Maddy, 2007, p. 2)

[10]Cf. (Papineau, 2015) for an in-depth discussion of different naturalistic positions in philosophy.

[11]"The Second Philosopher is a development of the naturalist in my (Maddy, 2001) and (Maddy, 2003) [...]; I adopt the name name here largely to avoid irrelevant debates about what 'naturalism' should be." (Maddy, 2007, p. 19, fn. 15)

As Maddy's terms suggest, the distinction is chiefly about whether philosophy or empirical observation have priority in the case of a conflict between the two. The first philosopher considers it possible that philosophical analysis results in a concept that, as an idealisation, may be in conflict with direct observations, and yet more worthy of philosophical study than the (pre-philosophical) everyday concept. In contrast, the second philosopher starts from the observation and would require a philosophical theory to explain it; a theory that does not match the observations would be scrutinised and possibly discarded:

> The theory of the real numbers, for example, is a fundamental component of the calculus and higher analysis, and as such is far more firmly supported than any philosophical theory of mathematical existence or knowledge. To sacrifice the former to preserve the latter is just bad methodology. (Maddy, 1990, p. 23)

We believe that the term *empirical philosophy* (or *empirically-based philosophy*) is best reserved for meta-philosophical positions that grant an important role for empirical data in the analysis of philosophical concepts and that would consider rejecting a philosophical theory if in clear conflict with empirical data. Note that this description does not specify the means by which these empirical data are collected. An empirical philosopher in this sense can espouse the experimental paradigm or reject it; he or she can consider qualitative methods of the social sciences or consider them too imprecise, etc.[12]

4.2 Empirical Methodology

Empirical data can be collected by several methods, ranging from unstructured observation via structured observation (using various methods) to experiments in controlled settings; some of the methods are qualitative, others quantitative.[13] The experimental method is so central for our contemporary idea of science that it requires constant reminders to recall that it is a relatively recent addition to the toolbox of the scientist and that it is only one among many empirical methods.[14] The relatively new field of "experimental philosophy" has mostly, but not exclusively, used the method of experiments. In the context of applying empirical findings to philosophy, a methodological discussion of the acceptable tools for data collection

[12]Note that in the above citation, Kauppinen (2007) only claims that "intuition statements [...] cannot [...] be tested through the methods of non-participatory social science", leaving it open whether empirical methods from the participatory social sciences might be able to serve as a test for intuition statements.

[13]We should like to emphasise that the experimental vs. non-experimental divide does not coincide with the quantitative vs. qualitative divide: most, but not all experimental work is quantitative and there are many quantitative non-experimental methods.

[14]The almost exclusive focus of philosophy of science on the experimental method has been criticized by some as an uncritical transfer of the method of modern physics to all of science (Lieberson and Lynn, 2002).

is very appropriate, especially since many of the philosophical applications involve empirical data from the social sciences where the choice of methodology is much more a matter of debate than in the exact sciences. Löwe and Van Kerkhove (in preparation) argue that a multiplicity of methods is particularly important in order to allow proper triangulation and that one should avoid relying solely on the experimental method. Obviously, it makes no sense to propose and change the name of the field *Experimental Philosophy* since it is well entrenched and accepted.[15]

4.3 The Source of Empirical Data

The final distinction is the one highlighted by Prinz: he calls the philosopher who uses empirical data from the literature an "empirical philosopher" and the philosopher who does the empirical work herself or himself an "experimental philosopher". Using a sartorial analogue, what Prinz calls "empirical philosophy" is using empirical data like *ready-to-wear* or *off-the-rack* clothes, provided in the shelves and racks of a store with no direct input by the customer expressing his or her preferences of clothing style or fit; on the other end of the spectrum, one would have the *bespoke* experience where the customer can determine every detail of the garment and the garment is then tailored exactly to the specifications of the customer and made to fit his or her body perfectly. Those who have the appropriate skills could even become the tailors themselves, getting rid of any need to communicate wishes and desires. *Bespoke* empirical philosophy would be a project in which the philosopher works very closely with the empirical scientist and designs an experiment or other observational activity jointly with her or him; the extreme case of *bespoke* would be *do-it-yourself* where the philosopher becomes an empirical scientist and does the empirical work herself or himself. Of course, the more extravagant and non-standard your desires and wishes with respect to your clothes are, the less likely it is that you will find these off the rack and you might have to move towards bespoke tailoring. Similarly, the empirical philosopher cannot expect that sociologists, education researchers, cognitive scientists, and historians work on matters relating to questions of philosophical relevance without being explicitly prompted to do so.

When Weinberg et al. (2001) started wondering about the culture-independence of judgments in Gettier-like situations, they became *do-it-yourself* empirical philosophers by doing the experiments related on in their paper and interpreting the empirical results in the epistemological context. However, let us consider the following alternative history: suppose there was a group of cognitive psychologists or linguists who were independently interested in the question whether people from different cultural backgrounds use the word "knowledge" in Gettier-type situations

[15]It has to be conceded that this more liberal usage of the term "experimental" is not unique to the field of experimental philosophy: not all of "experimental physics" is strictly speaking experimental in methodology.

differently and that this group produced the data from (Weinberg et al., 2001). Then Weinberg, Nichols and Stich could have published their paper as *ready-to-wear* empirical philosophy making the very same philosophical claim.

This *Gedankenexperiment* shows that whether the philosophical claim was made as *ready-to-wear* empirical philosophy or as *bespoke* empirical philosophy does not matter in principle for the quality of the philosophical argument. But, as in the sartorial word, not everything you need for your philosophical argument is available off the rack, forcing you to go bespoke or accept compromises in terms of fit between your data and the philosophical argument. The more you compromise on fit, the more it affects the quality of the philosophical argument; consequently, *ready-to-wear* empirical philosophy can only provide good arguments in a field where empirical data are sufficiently available to give the philosopher the resources to work with.

As in the case of tailoring, it is not universally the case that *bespoke* is better than *ready-to-wear*: if you find a high-quality garment off the rack that fits you very well, it may be considerable better than a bespoke garment from a mediocre tailor or (in the case of most of us with no expertise in tailoring) a garment that you made yourself. Similarly, in empirical philosophy, there are advantages and disadvantages to both *ready-to-wear* and *bespoke* approaches. In the first instance, philosophers are not empirical scientists, so *do-it-yourself* empirical philosophy requires that the philosopher acquire the skills and learn the techniques of another discipline, wasting time and energy that could be spent on something that they are more qualified for (such as doing philosophy) and possibly even leading to sub-standard or flawed empirical work. So, if the empirical data needed for a philosophical argument exist in the literature and fit the argument well, then *ready-to-wear* empirical philosophy based on these data might be preferable to *bespoke* or *do-it-yourself* empirical philosophy.

An example of an area where adequate and appropriate data are available is philosophy of mind: a rich literature of empirical research is published by cognitive scientists, psychologists, and neuroscientists on human cognition and its interaction with the mind and the brain, allowing philosophers of mind to do *ready-to-wear* empirical philosophy of mind without starting collaborative projects with said cognitive scientists, psychologists and neuroscientists.[16]

5 The Availability of Data

In Sect. 3, we have emphasized that it is chiefly the intended questions that distinguish *philosophy of mathematical practice* from the wider *study of mathematical cultures and practices*. In Sect. 4, we introduced the (re-named) Prinzian distinction

[16]This is not to say that there is no *bespoke* empirical philosophy of mind; as examples, let us mention the collaboration of Newen and Vogeley (e.g., David et al., 2008, 2006; Kockler et al., 2010) or the collaboration of van Lambalgen with cognitive neuroscientists (e.g., Baggio et al., 2008; Pijnacker et al., 2011).

of *ready-to-wear* and *bespoke* empirical philosophy. In this terminology, *ready-to-wear* philosophers of mathematical practice would rely on published data of people who work on mathematical practices and cultures to do their philosophical arguments, whereas philosophers of mathematical practice going *bespoke* would be more genuinely involved in the wider community studying mathematical practices and cultures in the form of joint research projects with researchers in mathematics education, anthropology, sociology, history, cognitive science and other fields.

We emphasised that the main factor in deciding whether you go bespoke or not is the availability of data in the literature. In light of this, it is very relevant for us that empirical data on mathematical practices and cultures are scarce, and this has been lamented by members of our community for the last decade.[17]

One field where we have a reasonable amount of data on mathematical research practices is the history of mathematics. In cognitive science, there is a rich literature on number cognition,[18] but as soon as we move to higher cognitive aspects of mathematical reasoning, there is not much research available. Of course, in mathematics education, we have a large and thriving literature using empirical methods, but most of it focuses on primary and secondary school education and very few researchers in mathematics education deal with tertiary education, research education or research itself.[19] In the case of sociology, Heintz (2000, p. 9) writes: "[d]ie Soziologie [begegnet] der Mathematik mit einer eigentümlichen Mischung aus Devotion und Desinteresse"; after Heintz's seminal book came out, a number of papers by members of our community have been published,[20] but compared to the sociology of other sciences, the literature is still very scarce.

This forces empirical philosophers of mathematics to do one of two things: remain *ready-to-wear* empirical philosophers of mathematics and restrict their attention to those questions that can be discussed with the scarce data available,

---

[17]The lament is already present in Kitcher and Aspray (1988, p. 17) in their introduction of the term "maverick tradition": "[B]ecause the [maverick] tradition is so recent, it now consists of a small number of scattered studies, studies that may not address the problems that are of most concern to mathematicians and historians." It can typically be found in the announcements of the events organised by our community; the following is a published version from (Löwe and Müller, 2010, p. vii): "[S]ociology of science mostly ignored mathematics presumably under the assumption that the human component of mathematical research is negligible."

[18]There is too much literature here to even give a few exemplary pointers; the two symposia *Mathematical Practice and Cognition* (organised by Alan Smaill, Markus Guhe, and Alison Pease) and *Mathematical Practice and Cognition II* (organised by Brendan Larvor and Alison Pease) at the 2010 and 2012 meetings of the Society for the Study of Artificial Intelligence and Simulation of Behaviour (AISB) in Leicester and Birmingham, respectively, got researchers in number cognition in touch with our community, and the special issue of the journal *Topics in Cognitive Science* (Volume 5, Issue 2, April 2013) containing the post-proceedings of the 2010 symposium shows the results of this cross-over nicely.

[19]Notable exceptions are, e.g., Weber and Mejia-Ramos (2011), Inglis and Alcock (2012), Inglis et al. (2013), and Weber et al. (2014).

[20]Cf., e.g., MacKenzie (2006), Greiffenhagen (2008), Rosental (2008), Greiffenhagen and Sharrock (2011a), Greiffenhagen and Sharrock (2011b), and Greiffenhagen (2014).

or become *bespoke* empirical philosophers of mathematics and start collaborative projects with the appropriate empirical scientists. Concerning the first option, historical empirical data on mathematical practices is much more available than empirical data from other neighbouring disciplines (such as sociology). We believe that this is one important factor in the perceived emphasis on historical studies in philosophy of mathematical practice.[21]

If a philosopher decides to do *bespoke* empirical philosophy of mathematics and to become actively involved in empirical research, it is important to notice that there is a large number of different disciplines involved in obtaining a multi-dimensional picture of mathematical practices and cultures. The number of involved disciplines is too large to hope that the philosopher could master all of the techniques from all of these disciplines; as a consequence, interdisciplinary collaboration with researchers from other fields is a necessary step for *bespoke* empirical philosophy of mathematics. This interdisciplinary collaboration in turn requires that the philosophers convince the researchers from other disciplines to get involved with their projects.

The following is an example of an effort to join forces with researchers from other disciplines by finding questions of relevance for everyone: In 2014 and 2015, the *International Union for History and Philosophy of Science and Technology* (IUHPST) ran a project *Cultures of Mathematical Research Training* funded by the *International Council of Science* (ICSU). This project brought researchers from all of the disciplines involved in our community together with representatives of funding agencies to produce a list of relevant research questions about the formation process of mathematical researchers that can be answered using empirical means. The project used a method for collaboratively identifying research priorities due to Sutherland et al. (2011). Since philosophy plays an important role in our community, philosophers were well represented during the two workshops of the project, guaranteeing that the philosophical legacy of our community does not get lost. Taking all of the represented disciplines into account, the project participants discussed which questions about the process of becoming a researcher in mathematics were the most relevant for the field as a whole. The resulting list of questions will be published as (Larvor and Löwe, 2016) and gives the result of this dialogue that required the philosophers to place their motivations into the larger context; it may serve as a catalyst for more interdisciplinary collaboration in the future.

[21]This "perceived emphasis on historical studies" is largely based on anecdotal eevidence and without a precise definition of what makes a study "historical" it is impossible to substantiate it. The following may serve as a rough indicator: at the first two APMP conferences (2010 in Brussels and 2013 in Urbana-Champaign), the percentage of abstracts explicitly mentioning the name of at least one pre-Second World War mathematician was 60 % (17 out of 28) and 52 % (13 out of 25), respectively.

6 Conclusion

We have argued that what we have called *our community* in Sect. 2 can be seen as two things: the sub-community of the wider field of *the study of mathematical practices and cultures* of those people motivated by philosophical questions and at the same time the sub-community of philosophy of mathematics that would be *empirical* in the sense of Sect. 4. The term "philosophy of mathematical practice" emphasises the second characterisation. We have also discussed that the number of researchers actively working on mathematical practices and cultures that would be of relevance for philosophers of mathematics is small, and therefore, members of our community cannot expect to do *ready-to-wear* empirical philosophy of mathematics and "work by citation"; instead, they have to rely on close collaborations with cognitive scientists, researchers in mathematics education, sociologists, anthropologists, psychologists, and representatives of many other disciplines. In practice, we need to get these people excited about our questions and convince them that it is worthwhile to collaborate with us on questions. We might wonder whether the emphasis on questions driven by traditional philosophy of mathematics could constitute a practical obstacle in this endeavour.

References

Baggio, G., van Lambalgen, M., & Hagoort, P. (2008). Computing and recomputing discourse models: An ERP study. *Journal of Memory and Language, 59*(1), 36–53.

Buldt, B., Löwe, B., & Müller, T. (2008). Towards a new epistemology of mathematics. *Erkenntnis, 68*, 309–329.

Corfield, D. (2003). *Towards a philosophy of real mathematics*. Cambridge: Cambridge University Press.

David, A., Aumann, C., Santos, N., Bewernick, B., Eickhoff, S., Newen, A., et al. (2008). Differential involvement of the posterior temporal cortes in mental versus spatial perspective taking. *Social Cognitive and Affective Neuroscience, 3*, 279–289.

David, N., Bewernick, B., Cohen, M., Newen, A., Lux, S., Fink, G., et al. (2006). The self-other distinction in social cognition—Perspective-taking and agency in a virtual ball-tossing game. *Journal of Cognitive Neuroscience, 18*, 898–910.

Dutilh Novaes, C. (2012). Towards a practice-based philosophy of logic: Formal languages as a case study. *Philosophia Scientiae, 16*, 71–102.

Ferreirós, J. (2014). *Pragmatism in the philosophy of mathematics*. Unpublished talk given at the meeting Pragmatism and the practical Turn in Philosophy of Science of the Académie Internationale de Philosophie des Sciences held in Pont-à-Mousson, September 10–14, 2014.

Ferreirós, J. (2015). *Mathematical knowledge and the interplay of practices*. Princeton: Princeton University Press.

Greiffenhagen, C. (2008). Video analysis of mathematical practice? Different attempts to 'open up' mathematics for sociological investigation. *Forum: Qualitative Social Research, 9*(3), 1–20.

Greiffenhagen, C., & Sharrock, W. (2011a). Does mathematics look certain in the front, but fallible in the back? *Social Studies of Science, 41*(6), 839–866.

Greiffenhagen, C., & Sharrock, W. (2011b). Sources for myths about mathematics. On the significance of the difference between finished mathematics and mathematics-in-the-making. In K. François, B. Löwe, T. Müller, & B. Van Kerkhove (Eds.), *Foundations of the formal sciences VII. Bringing together philosophy and sociology of science*. Studies in logic (Vol. 32, pp. 91–109). London: College Publications.

Greiffenhagen, C. W. K. (2014). The materiality of mathematics: Presenting mathematics at the blackboard. *British Journal of Sociology, 65*(3), 502–528.

Heintz, B. (2000). *Die Innenwelt der Mathematik. Zur Kultur und Praxis einer beweisenden Disziplin* (pp. 1–22). Wien: Springer.

Hersh, R. (1997). *What is mathematics, really?* Oxford: Oxford University Press.

Inglis, M., & Alcock, L. (2012). Expert and novice approaches to reading mathematical proofs. *Journal for Research in Mathematics Education, 43*, 358–390.

Inglis, M., Mejia-Ramos, J. P., Weber, K., & Alcock, L. (2013). On mathematicians' different standards when evaluating elementary proofs. *Topics in Cognitive Science, 5*(2), 270–282.

Jullien, C., & Soler, L. (2014). Conceptions of mathematical practices: Some remarks. Commentary on "The impact of the philosophy of mathematical practice on the philosophy of mathematics", by Jean Paul Van Bendegem. In L. Soler, S. Zwart, M. Lynch, & V. Israel-Jost (Eds.), *Science after the practice turn in the philosophy, history, and social studies of science*. Routledge studies in the philosophy of science (pp. 227–237). New York: Routledge.

Kauppinen, A. (2007). The rise and fall of experimental philosophy. *Philosophical Explorations, 10*(2), 95–118.

Kitcher, P., & Aspray, W. (1988). An opinionated introduction. In W. Aspray & P. Kitcher (Eds.), *History and philosophy of modern mathematics*. Minnesota studies in the philosophy of science (Vol. XI, pp. 3–57). Minneapolis: University of Minnesota Press.

Kockler, H., Scheef, L., Tepest, R., David, N., Bewernick, B. H., Newen, A., et al. (2010). Visuospatial perspective taking in a dynamic environment: Perceiving moving objects from a first-person-perspective induces a disposition to act. *Consciousness and Cognition, 19*(3), 690–701.

Larvor, B. P. (2016). What are mathematical cultures? In S. Ju, B. Löwe, T. Müller, & Y. Xie (Eds.), *Cultures of Mathematics and Logic, Selected Papers from the Conference in Guangzhou*, China, November 9–12, 2012. Trends in the history of science. Berlin: Springer, pp. 1–22.

Larvor, B. P., & Löwe, B. (2016, in preparation). Cultures of mathematical research training. White Paper reporting on a project organised by the International Union for History and Philosophy of Science and Technology.

Lieberson, S., & Lynn, F. B. (2002). Barking up the wrong branch: Scientific alternatives to the current model of sociological science. *Annual Review of Sociology, 28*, 1–19.

Löwe, B. (2014). Mathematics and the new technologies. Part I: Philosophical relevance of a changing culture of mathematics. In P. Schroeder-Heister, G. Heinzmann, W. Hodges, & P. E. Bour (Eds.), *Logic, Methodology and Philosophy of Science, Proceedings of the 14th International Congress (Nancy), Logic and Science Facing the New Technologies* (pp. 399–407). London: College Publications.

Löwe, B., & Müller, T. (Eds.). (2010). *PhiMSAMP. Philosophy of mathematics: Sociological aspects and mathematical practice*. Texts in philosophy (Vol. 11). London: College Publications.

Löwe, B., Müller, T., & Müller-Hill, E. (2010). Mathematical knowledge: A case study in empirical philosophy of mathematics. In B. Van Kerkhove, J. De Vuyst, & J. P. Van Bendegem (Eds.), *Philosophical perspectives on mathematical practice*. Texts in philosophy (Vol. 12, pp. 185–203). London: College Publications.

Löwe, B., & Van Kerkhove, B. (in preparation). Methodological triangulation in empirical philosophy of mathematics.

MacKenzie, D. (2006). Computers and the sociology of mathematical proof. In R. Hersh (Ed.), *18 unconventional essays on the nature of mathematics* (pp. 128–146). Berlin: Springer.

Maddirala, N. (2014). *Philosophy of logical practice: A case study in formal semantics* (Master's thesis). Universiteit van Amsterdam. ILLC Publications MoL-2014-15.

Maddy, P. (1990). *Realism in mathematics*. Oxford: Oxford University Press.

Maddy, P. (1997). *Naturalism in mathematics*. Oxford: Oxford University Press.
Maddy, P. (2001). Naturalism: Friends and foes. In J. Tomberlin (Ed.), *Philosophical perspectives, Volume 15: Metaphysics* (pp. 37–67). Oxford: Blackwell.
Maddy, P. (2003). Second philosophy. *Journal of the Indian Council of Philosophical Research, 20*, 73–106.
Maddy, P. (2007). *Second philosophy. A naturalistic method*. Oxford: Oxford University Press.
Mancosu, P. (2008a). Introduction. In P. Mancosu (Ed.), *The philosophy of mathematical practice* (pp. 1–21). Oxford: Oxford University Press.
Mancosu, P. (Ed.). (2008b). *The philosophy of mathematical practice*. Oxford: Oxford University Press.
Marrou, H.-I. (1934). "*Doctrina*" et "*Disciplina*" dans la langue des pères de l'église. *Bulletin du Cange, Archivum Latinitatis Medii Aevi, 9*, 5–25.
Müller-Hill, E. (2009). Formalizability and knowledge ascriptions in mathematical practice. *Philosophia Scientiae, 13*(2), 21–43.
Müller-Hill, E. (2011). *Die epistemische Rolle formalisierbarer mathematischer Beweise—Formalisierbarkeitsorientierte Konzeptionen mathematischen Wissens und mathematischer Rechtfertigung innerhalb einer sozio-empirisch informierten Erkenntnistheorie der Mathematik* (PhD thesis). Rheinische Friedrich-Wilhelms-Universität Bonn.
Papineau, D. (2015). Naturalism. In E. N. Zalta (Ed.), *The Stanford encyclopedia of philosophy* (Fall 2015 Edn.). CSLI, Stanford.
Pijnacker, J., Geurts, B., van Lambalgen, M., Buitelaar, J., & Hagoort, P. (2011). Reasoning with exceptions: An event-related brain potentials study. *Journal of Cognitive Neuroscience, 23*(2), 471–480.
Prinz, J. J. (2008). Empirical philosophy and experimental philosophy. In J. Knobe & S. Nichols (Eds.), *Experimental philosophy* (pp. 198–208). Oxford: Oxford University Press.
Rosental, C. (2008). *Weaving self-evidence. A sociology of logic*. Princeton: Princeton University Press.
Stokhof, M., & van Lambalgen, M. (2011). Abstractions and idealisations: The construction of modern linguistics. *Theoretical Linguistics, 37*(1–2), 1–26.
Sutherland, W. J., Fleishman, E., Mascia, M. B., Pretty, J., & Rudd, M. A. (2011). Methods for collaboratively identifying research priorities and emerging issues in science and policy. *Methods in Ecology and Evolution, 2*(3), 238–247.
Van Bendegem, J. P. (2014). The impact of the philosophy of mathematical practice on the philosophy of mathematics. In L. Soler, S. Zwart, M. Lynch, & V. Israel-Jost (Eds.), *Science after the practice turn in the philosophy, history, and social studies of science*. Routledge studies in the philosophy of science (pp. 215–226). London: Routledge.
Van Kerkhove, B. (Ed.). (2008). *New Perspectives on Mathematical Practices. Essays in Philosophy and History of Mathematics*, Brussels, Belgium, March 26–28, 2007. Singapore: World Scientific.
Van Kerkhove, B., De Vuyst, J., & Van Bendegem, J. P. (Eds.). (2010). *Philosophical perspectives on mathematical practice*. Texts in philosophy (Vol. 12). London: College Publications.
Van Kerkhove, B., & Van Bendegem, J. P. (Eds.). (2007). *Perspectives on mathematical practices. Bringing together philosophy of mathematics, sociology of mathematics, and mathematics education*. Logic, epistemology, and the unity of science (Vol. 5). Berlin: Springer.
Weber, K., Inglis, M., & Mejia-Ramos, J. P. (2014). How mathematicians obtain conviction: Implications for mathematics instruction and research on epistemic cognition. *Educational Psychologist, 49*, 36–58.
Weber, K., & Mejia-Ramos, J. P. (2011). Why and how mathematicians read proofs: An exploratory study. *Educational Studies in Mathematics, 76*(3), 329–344.
Weinberg, J. M., Nichols, S., & Stich, S. (2001). Normativity and epistemic intuitions. *Philosophical Topics, 29*, 429–460.

Abacus as a Programming Language: Computer Science and History of Mathematics

Baptiste Mélès

Abstract

I show in this paper that programming paradigms, such as imperative or functional programming, can be put in correspondence with computing styles of calculating tools. The Chinese abacus actually follows a functional programming paradigm, while the ancient Chinese counting rods display an imperative programming style. More generally, I show that abstract concepts, such as currying or the semantics of programming languages, can be transposed from theoretical computer science to the historical description of the practice of computing tools.

1 Introduction

Computing tools, such as abaci, logarithm tables, slide rules and counting on paper, differ both in physical structure and calculating capability. Computability theory and computer science may help historians of mathematics shed light on the latter aspect.

As a matter of fact, every computing tool comes with an implied set of operations and algorithms, which can be learned one after another by the practitioner: addition and multiplication on the abacus, exponentiation on the slide rule, resolution of systems of linear equations with the Chinese counting rods... Each of those tools also seems to have internal limitations: there is apparently no easy way of performing additions with a slide rule, and no obvious way to compute logarithms with an abacus. Calculating tools do not all have the same computational power.

B. Mélès (✉)
Laboratoire d'Histoire des Sciences et de Philosophie—Archives Henri-Poincaré, Unité Mixte de Recherche du Centre National de Recherche Scientifique et de l'Université de Lorraine (UMR 7117), 91 avenue de la Libération, BP 454, 54001 Nancy, France
e-mail: baptiste.meles@univ-lorraine.fr

S. Ju et al. (eds.), *Cultures of Mathematics and Logic*, Trends in the History of Science, DOI 10.1007/978-3-319-31502-7_3

What may be more surprising is that, even in the case of two computing tools being able to compute the *same* operation, e.g., addition, they may have different ways of calculating it, depending on the arithmetical properties they algorithmically handle, as though they were thereby carrying different "interpretations" of this operation. Is there a way to describe with full generality how given computing tools "represent" algorithms?

This is a question both for historians of mathematics and computer scientists. Indeed, one way to answer this question is to consult the classic works that discuss the computational in question, such as Fibonacci's *Liber Abaci*[1] for counting on paper or Cheng Dawei's *Suanfa Tongzong*[2] for the abacus. An other way of answering this question is to consider computing tools as machines, as Turing first did in his article "On Computable Numbers:"[3] the definition of his machines comes from the description of a working *human* "computer." Our aim is to show how both approaches, history of mathematics and theoretical computer science, can be fruitfully combined. There is no reason to be afraid: using concepts of computer science for cultures which had no computers is no more an anachronism than speaking of "place-value notation" for cultures which certainly had a use, but no concept, of "place-value notation."

We show in this paper that programming paradigms, such as imperative or functional programming, can be put in correspondence with computing styles of calculating tools. The Chinese abacus actually follows a functional programming paradigm, while the ancient Chinese counting rods display an imperative programming style. More generally, we intend to show that abstract concepts, such as currying or the semantics of programming languages, can be transposed from theoretical computer science to the historical description of the practice of computing tools.

2 Counting Rods as an Imperative Language

2.1 Data Structure and Variable Assignment

Karine Chemla, in the French edition of the Chinese classical book *The Nine Chapters on Mathematical Procedures*,[4] effectively used computer science to describe algorithmic aspects of ancient Chinese mathematics.

The computing tool the *Nine Chapters* allude to is a board on which rods denoting digits can be laid out and moved.[5] This board may be seen as a two-dimensional array of digits, in which each dimension, be it vertical or horizontal,

[1](Fibonacci, 2002).

[2](Cheng Dawei 程大位, 1993).

[3](Turing, 1936).

[4](Chemla and Guo Shuchun, 2004).

[5](Chemla, 1996).

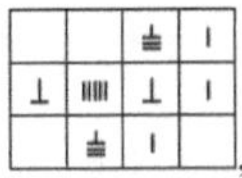

, i.e.,

		8	1
6	5	6	1
	8	1	

Fig. 1 Final step of a multiplication with the counting rods (*Sunzi Suanjing*, I.16)

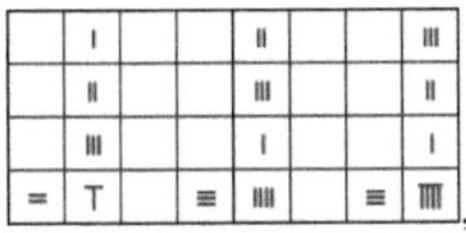

, i.e.,

	1			2			3
	2			3			2
	3			1			1
2	6		3	4		3	9

Fig. 2 A system of linear equations (*Nine Chapters*, VIII.1)

has its usefulness. There is first a use of the *horizontal* layout, which simply consists in "writing down" the digits composing the number, such as

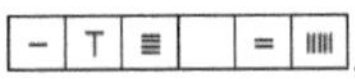

, *i.e.*,

1	6	4		2	5

There is secondly a *vertical* layout of numbers, which can be used in some algorithms to align numbers, or parts of numbers, e.g., when multiplying numbers as in the *Sunzi Suanjing* (Fig. 1).[6]

There is, lastly, a second kind of *horizontal* layout of numbers, which allows to connect somehow correlated numbers, as for systems of linear equations (cf. Fig. 2).

This bi-dimensional array which characterizes the Chinese counting board could by analogy be called its *data structure*: in computer science, data structures—such as registers, queues, stacks and arrays—determine how objects, typically numbers, are encoded within memory and what elementary relations hold between them.[7]

But there is more. As Karine Chemla showed, the vocabulary of computer science can be fruitfully used to describe how the algorithms are performed. In particular, "variable assignment," which Knuth calls "the all-important operation,"[8] is applied on the counting board, for square roots as well as for divisions.[9] For example, the algorithm for square roots using "borrowed rods" (*jie suan* 借算), as described in the *Nine Chapters* 4.16, requires to put down some intermediary results "as auxiliaries" (*fu* 副) (cf. Fig. 3).[10] During the execution of the algorithm, one indeed has to assign a value to this "auxiliary" or intermediary variable, which belongs neither to the operands nor to the result, and is merely dropped when the algorithm halts. This is why Karine Chemla argues that the algorithms of

[6](Sunzi 孫子, 1993).

[7]For detailed examples of data structures, cf. (Knuth, 1997, Chapter 2).

[8](Knuth, 1997, p. 3).

[9](Chemla and Guo Shuchun, 2004, pp. 25, 110, & 324–327).

[10]See also the *Nine Chapters*, 1.6: to simplify parts, one first needs to compute "in auxiliary" (*fu* 副) the greatest common divisor.

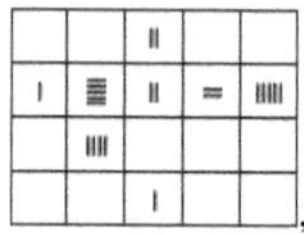

, i.e.,

		2			quotient
1	5	2	2	5	dividend
	4				divisor
		1			auxiliary

Fig. 3 Variable assignment in the square root algorithm (*Nine Chapters*, IV.12)

the *Nine Chapters* can be described in terms of lists of operations using variable assignment.[11]

2.2 Imperative Languages

The hypotheses of this argument actually deserve some attention.

Karine Chemla actually relies on Donald Knuth's classic book *The Art of Computer Programming*, in which all algorithms are designed for an abstract machine called `MIX`.[12] This is a von Neumann machine: it contains registers and memory addresses which allow us to store, not only input and output numbers, but also auxiliary numbers, making variable assignment possible.[13] Moreover, Knuth's programs are written, not as mathematical compositions of functions, but as *sequences* of instructions operating on the variables, i.e., ultimately on the registers; every instruction can be somehow disconnected from the previous and the next ones, which can manipulate different variables: we must not always use the results of the previous instruction as input data of the current instruction. This allows Knuth to formalize every algorithm in his own assembly language,[14] which is an *imperative* language, i.e., precisely a language which extensively uses variable assignment, and where algorithms are described as mere sequences of instructions. Thus, the properties of Knuth's imperative language, which relies on a von Neumann machine, fit particularly well the computing style of the counting board described by Karine Chemla.

Algorithms for counting on paper, by the way, display the same general computational properties: they are sequences of operations with variable assignments. This will not be surprising if we consider that this computing tool precisely was Turing's model of inspiration,[15] and that Turing machines explicitly inspired von Neumann machines, hence imperative languages.[16]

This family resemblance between given computing tools and an abstract machine, and hence between computing tools and a programming paradigm, leads

[11](Chemla and Guo Shuchun, 2004, pp. 21–25).

[12](Knuth, 1997, p. 124–144).

[13](von Neumann, 1993).

[14](Knuth, 1997, p. 144–163).

[15](Turing, 1936, §1 and 9).

[16](Backus, 1978, p. 615–616).

us naturally to the following question: since there exist a lot of programming paradigms which are not imperative, are there computing tools corresponding to some of them?

There actually are functional languages like Lisp and Haskell, declarative languages like Prolog, object-oriented languages like C++, etc. Although most available languages are "Turing-equivalent," i.e., enjoy the same computational power, they display syntactic differences, which express as many various ways of representing to oneself what computation and its objets are. As Peter Rechenberg argued in 1990, programming paradigms express "thought models."[17] Should we think that all computing tools can be reduced to one and the same model, Knuth's imperative style, or are there other styles of calculating tools, corresponding to other styles of programming languages?

The example of the Chinese abacus will show that there are calculating styles which are not imperative, and that these stylistic variations can have a direct influence on the pedagogical styles and strategies.

3 Abacus as a Functional Language

3.1 Data Structure of the Abacus

There are several kinds of abacus. Greek and Roman abaci are boards used with little stones,[18] while the Chinese abacus (*suanpan* 算盤), the Japanese abacus (*soroban* 算盤) and the Russian abacus (*schoty* счёты) consist of rods and beads.

Typically, the Chinese abacus bears five unary beads and two quinary beads on each rod (cf. Fig. 4a); the Japanese abacus, four unary beads and one quinary bead on each rod (cf. Fig. 4c); the Russian abacus, ten unary beads on most rods, but only four unary beads on a rod dedicated to quarters of ruble (cf. Fig. 4b). In this paper, we shall focus on the Chinese abacus. Our illustrations will represent it with 13 rods. Whilst some of our results can be directly applied to other kinds of abaci than the *suanpan*, others can not.

What is the data structure of the Chinese abacus? In the simplest case, the *suanpan* can be used to encode *one* number. This number typically is the *first* number of the calculus, say the first term of an addition; but it can also be the *final* result of the computation, which is shown by the merchant to the client. Figure 5 shows how to encode the classical number 123 456 789, using unary beads on the bottom and quinary beads on the top of the central bar.

Some algorithms require one to encode at once *two* different numbers on the abacus. For instance, whoever wants to compute the greatest common divisor of 49 and 91 just has to encode each number on one side of the abacus, and to remove the least one from the greatest, until both are the same (Fig. 6).

---

[17](Rechenberg, 1990).

[18](Schärlig, 2001).

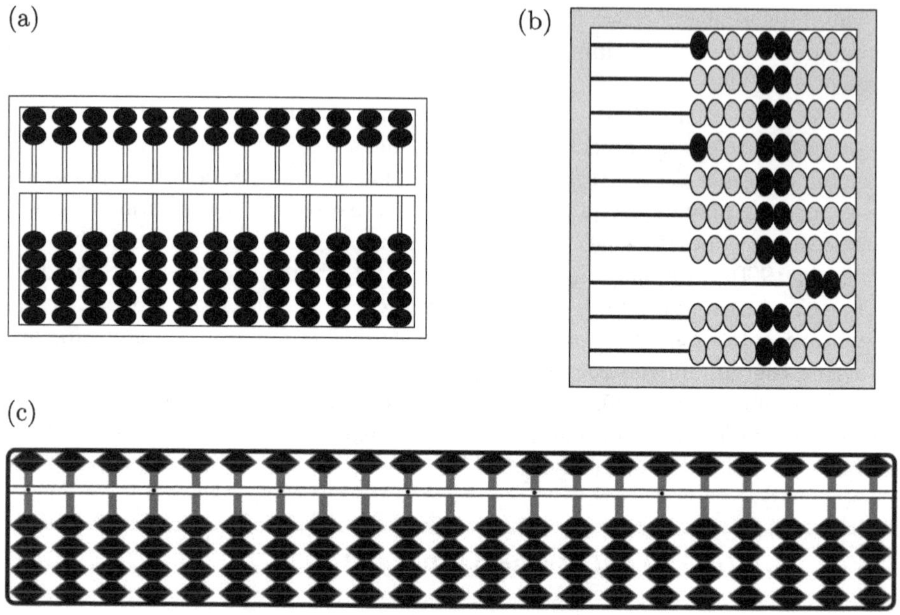

Fig. 4 Various abaci. (**a**) A Chinese abacus (*suanpan*). (**b**) A Russian abacus (*schoty*). (**c**) A Japanese abacus (*soroban*)

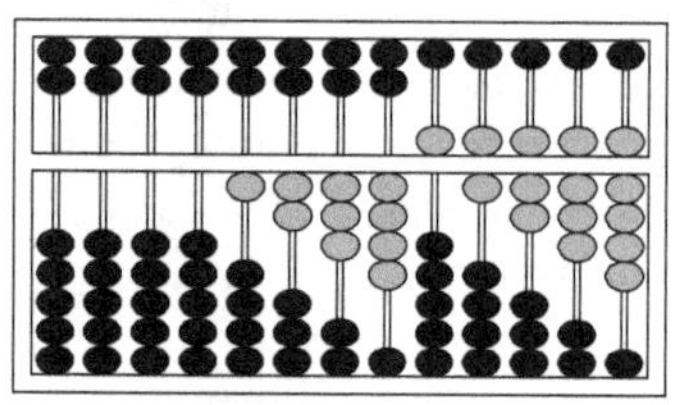

Fig. 5 The number 123 456 789

For some algorithms, one even needs to encode *three* numbers. This is the case for the Euclidean algorithm (cf. Fig. 7).

Contrary to the data structure of the Chinese counting board, which consists in a full array of number, the data structure of a *suanpan* only amounts to a sequence of a few numbers, which are the only "objects" encoded on the calculating tool. In most cases—addition, subtraction, multiplication, division...—the algorithms for the abacus will only require the encoding of *one* visible number. This may be surprising, for most of our operations, binary as they are, involve three numbers, namely two operands and one result. We must therefore understand what transformations binary operations must be subjected to get implemented on the abacus.

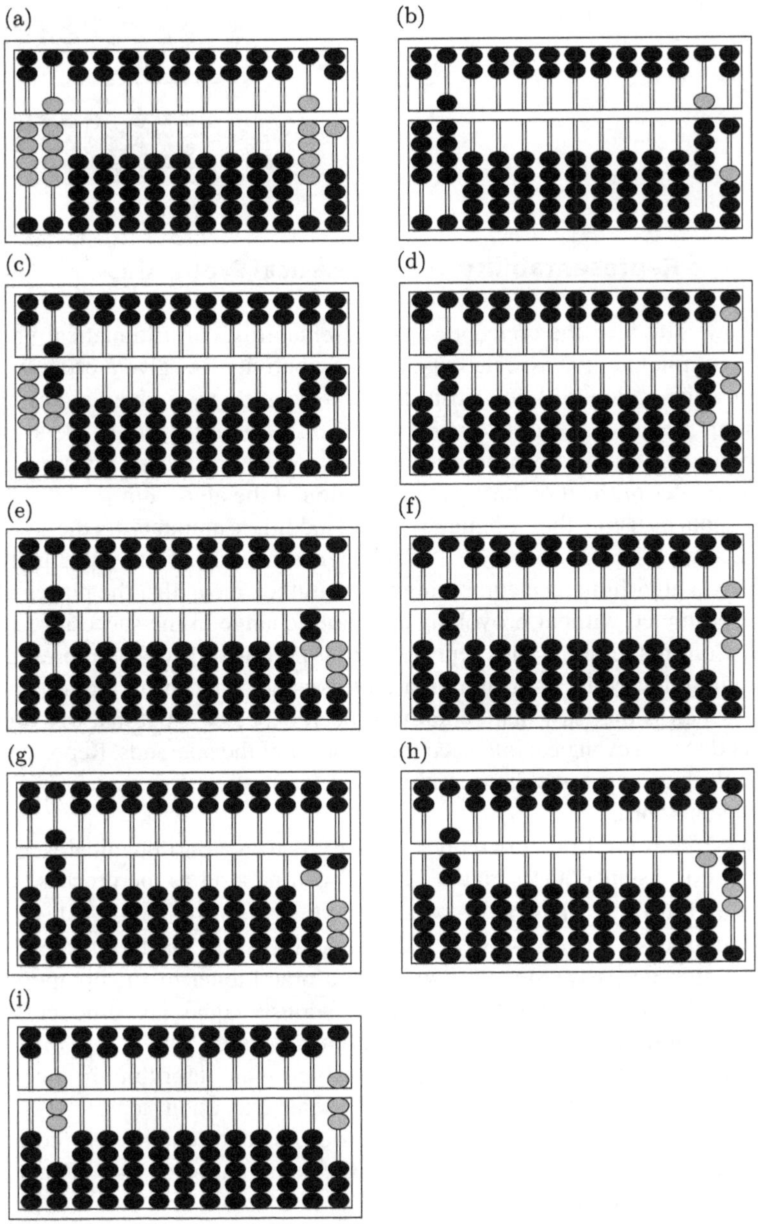

Fig. 6 How to compute the greatest common divisor of 49 and 91. (**a**) Encode 49 and 91. (**b**) Remove 49 from 91 (91 − 50 + 1). (**c**) Remove 42 from 49 (49 − 40 − 2). (**d**) Remove 7 from 42 (42 − 10 + 5 − 2). (**e**) Remove 7 from 35 (35 − 10 + 3). (**f**) Remove 7 from 28 (28 − 7). (**g**) Remove 7 from 21 (21 − 10 + 3). (**h**) Remove 7 from 14 (14 − 10 + 5 − 2). (**i**) Both numbers are equal: the GCD is 7

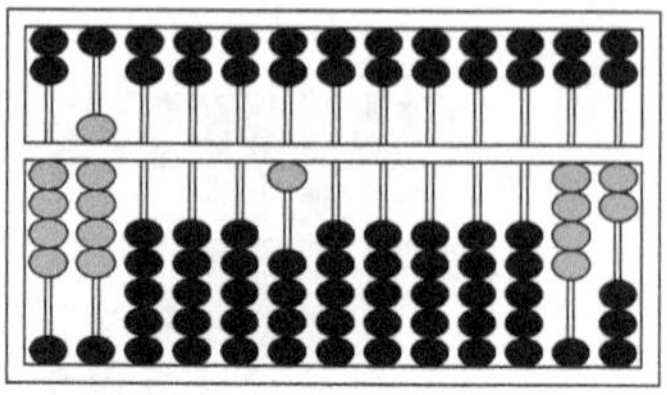

Fig. 7 Euclidean division: $91 = 1 \times 49 + 42$

3.2 The Representability of Arithmetical Properties

Let us now introduce the concept of "representability" of arithmetical properties such as commutativity, associativity or distributivity. A given property of an arithmetical operation will be said to be *representable* by a given algorithm on a given computing tool if the invariance of the result with respect to the syntactical transformations it allows can be put in correspondence with an invariance of the successive states of the tool during the execution of the algorithm.

For instance, using the commutativity of addition amounts to the syntactical transformation of $a + b$ into $b + a$. The *representability* of the commutativity of addition by a algorithm on a computing tool will then mean that the places of a and b can be permuted without provoking any other change in the successive states of the tool. Commutativity is thus represented by permutability. The representability of commutativity by an algorithm on a tool does not only mean that this property is true—which is the least that could have been expected—but that this algorithm moreover does not even take into account the order of the operands. Representability in general thus expresses how arithmetical properties are made "visible" by algorithms on tools.

An easy criterium to define representability is to read arithmetical formulae as programs. For instance, the formula $11 + \frac{14}{2}$ can be read as the process "take 11, and then add to it the division of 14 by 2," while the formula $\frac{14}{2} + 11$ would be read as the process "divide 14 by 2, and then add 11 to it," each of those programs describing the successive steps followed by a practitioner of the computing tool. *Arithmetically* speaking, both formulae are obviously equal, i.e., produce the same result. The question is to know whether both processes also are *algorithmically* equal. The arithmetical equality is representable by an algorithm on a computing tool if the corresponding programs follow the same execution steps.

In this sense, the commutativity of addition is representable by the common algorithm on paper. Let us actually compare the computation of $14+27$ and $27+14$ using the same algorithm:

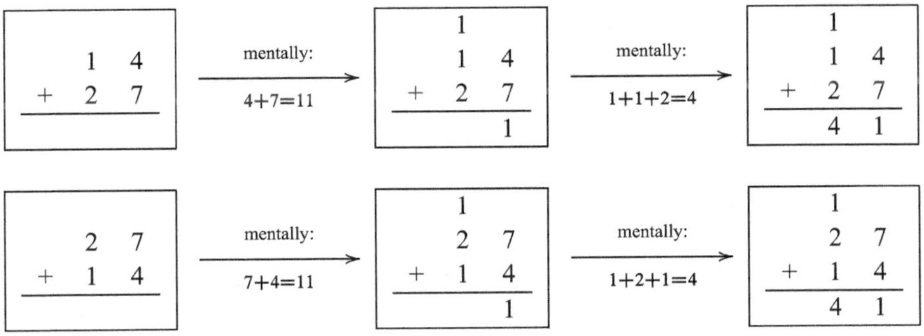

Be commutativity present or not in the practitioner's head, it is for sure invisible on the very tool.

By contrast, the commutativity of multiplication is not representable on paper by the common algorithm, for the computations follow different steps:

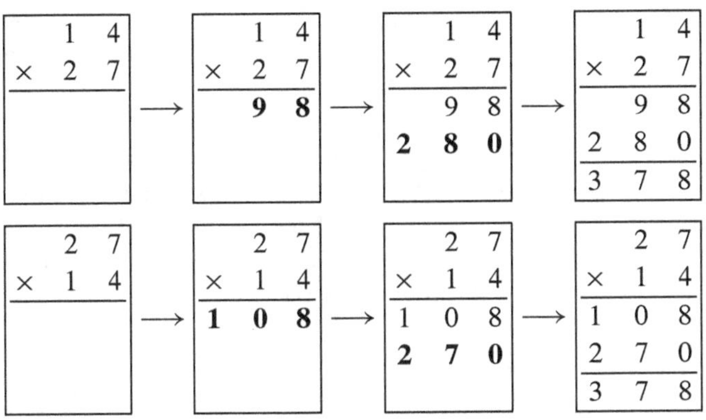

Despite the law of distributivity, the "programs" $(7 \times 14) + (20 \times 14)$ and $(4 \times 27) + (10 \times 27)$—which obviously produce the same result—do not follow the same execution steps on paper. By contrast, the illustration of multiplication by rectangles naturally allows us to represent the commutativity of multiplication[19]:

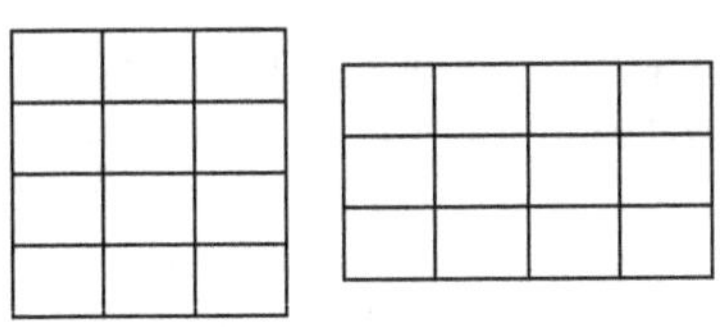

That the commutativity of addition, and not of multiplication, is representable on paper, may be the reason why we are so accustomed to distinguish between a multiplicand and a multiplier, but not so much between an "addend" and an

[19]Many other properties could be illustrated using the same diagrams: associativity and commutativity of addition, associativity and distributivity of multiplication.

"adder." The commutativity of operations does not imply that their operands play symmetrical roles during the computation. More generally, true arithmetical properties may not be represented on computing tools: such truths are not always visible. Arithmetical properties are not always converted into algorithmic properties.

3.3 Currying Operations

On the abacus, unlike by counting on paper, even the commutativity of addition is not representable: the computations of 14 + 27 and 27 + 14 follow different steps (cf. the details on Fig. 8). Under such circumstances, the distinction between an addend and an adder seems highly appropriate. How is it possible?

When we count on paper, the operation of addition is somehow external to the operands. Looking at our addition table, we first have to find the two operands, and the intersection of their respective line and column in the table does not "come" from any of the numbers more than from the other: it "comes" from both at once. When we write 12 + 81, we do not always pay attention to the issue of knowing whether we have to add 12 to 81, or 81 to 12. Both operands are objects; the act of operation comes from outside.

The case is very different for the practitioner of the abacus, as a very simple example will show. Let us indeed compute 12 + 81 on the abacus (Fig. 9). This very example suffices to show that the two operands of an addition do not play symmetrical roles. At the very beginning of the execution of our algorithm, when the abacus is still empty, the first operand of the addition, 12, is used, not really as a "number," i.e., as a passive object, but as the part of an act, of a transition operation from one state of the abacus to another. This number, 12, is then "frozen" as an object, a static state of the abacus, while the second one, 81, becomes in turn a part of a new operation, in which 12 is now passive, while 81 is active. Once this addition has been performed, its result, 93, can become a passive object for new additions, and so forth. Actually, we do not really compute on the abacus what we could write in modern notation as "12 + 81:" we apply the operation "+81" to the object "12." Unlike the operands of an addition on paper, which are both objects subjected to a common act, operands here play asymmetrical roles: one is an object, the other is an act.

This asymmetry between the operands of an operation does not only exist for the addition, but also for all classical algorithms on the abacus, such as multiplication, division, square root, etc. This is a fundamental feature of computing with an abacus.

This transformation of a commutative operation into an asymmetrical algorithm is what logicians and computer scientists call "currying." Schönfinkel, before Curry, discovered that every *n*-ary function could be transformed into a composition of unary functions, provided that their respective "values" can in turn be functions.[20] If one does not curry the operation, 12 + 81 just consists in taking two numbers,

[20](Schönfinkel, 1924, § 2).

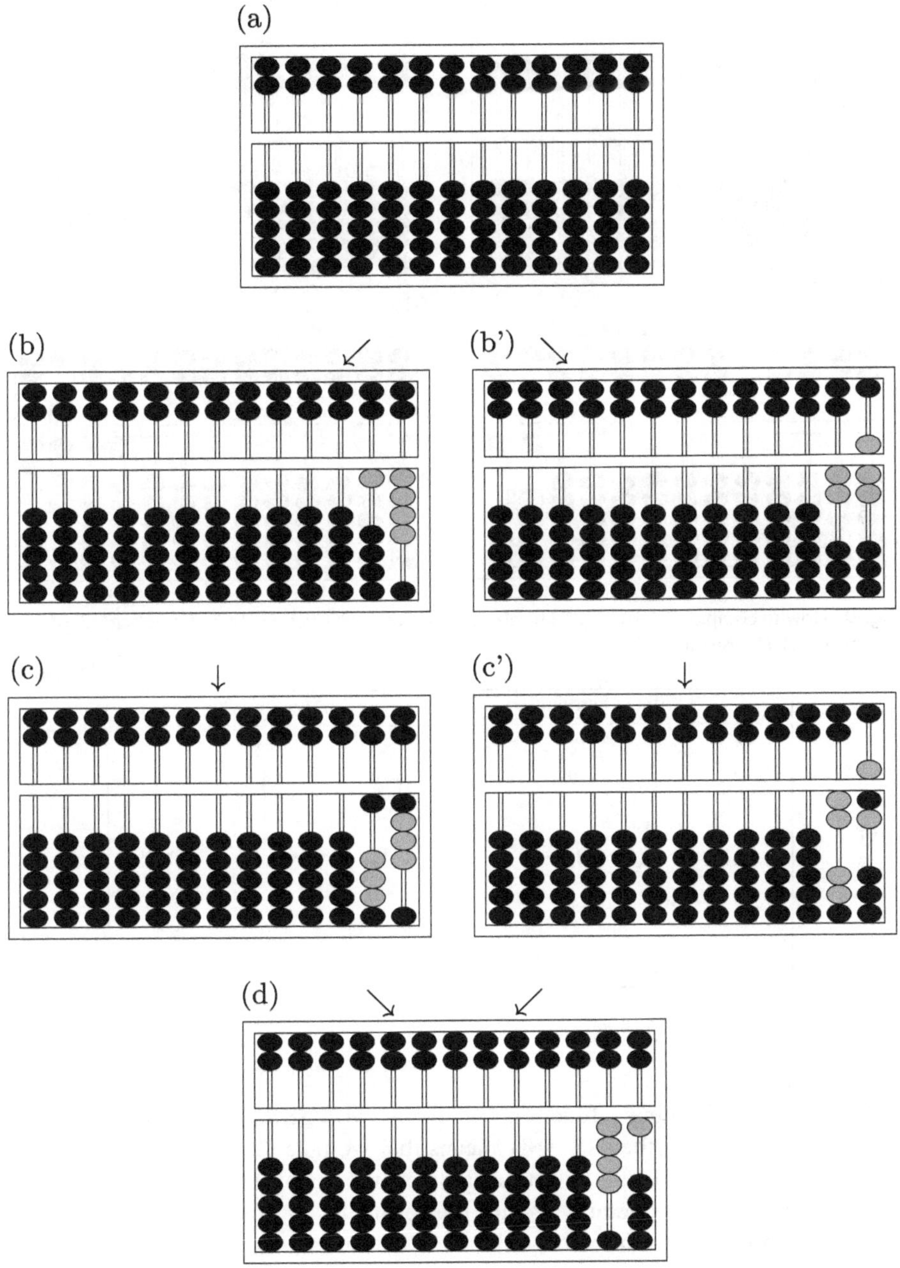

Fig. 8 14 + 27 = 27 + 14 (through different computations). (**a**) Reset the abacus. (**b**) Encode 14 as operand. (**b′**) Encode 27 as operand. (**c**) Add 27 (i.e., +20 + (10 − 3)). (**c′**) Add 14 (i.e., +10 + (10 − 6)). (**d**) The result is 41

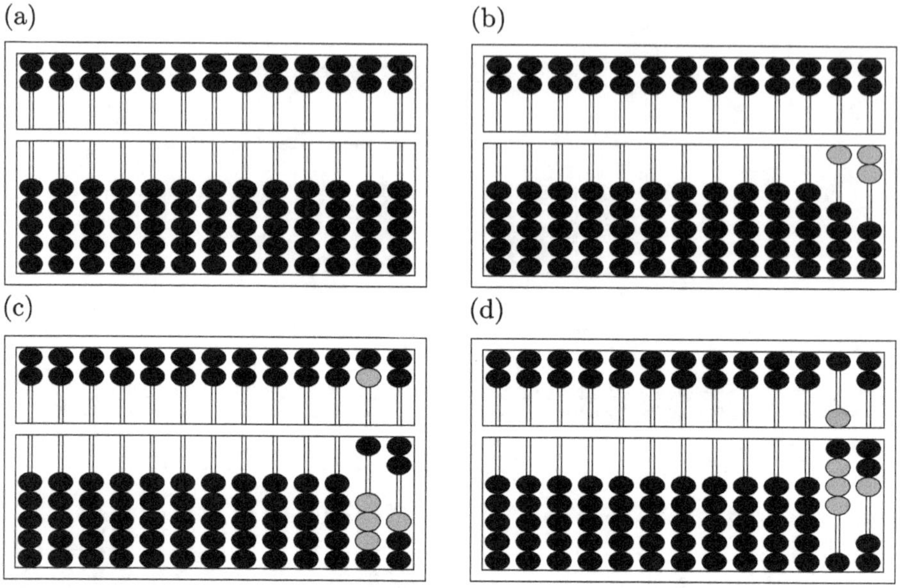

Fig. 9 How to compute 12 + 81. (**a**) The abacus is empty. (**b**) Encode the first number: 12. (**c**) Add 80 and 1. (**d**) The result is 93

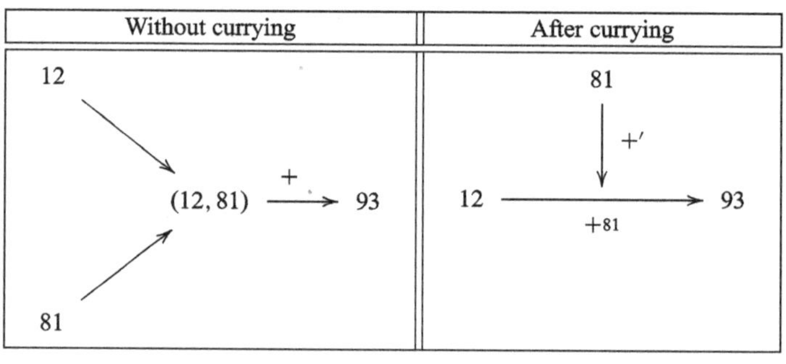

Fig. 10 The currying of 12 + 81

and in applying the binary function "+" to its two operands (Fig. 10). But if one curries this operation, then 12 + 81 means that we take 12 as an object (a proper operand), transform 81 into an act (such as operations of the hand on the abacus), which transforms the operand 12 into the result 93. In this latter example, 81 is no longer an operand—an object—for the addition, but a part of the operation of addition itself.

This explains the fact that the commutativity of addition is not representable on the abacus: both operands do not play symmetrical roles, since one is an object while the other belongs to an act.

This fact has other direct consequences. In particular, there is no "variable assignment" on the abacus. The operand and the result actually do not occupy distinct places: the final result is nothing but a transformation of the input data themselves, without creating intermediary nor auxiliary values. Another consequence is that an algorithm, on the abacus, is not a mere sequence, but a *composition of functions*: each step of the algorithm takes the actual state of the abacus and transforms it into a new state, which will be the starting point of the next step.

Those two properties—the lack of variable assignment, and the limitation of algorithms to composition of functions—define, following John Backus,[21] the *functional style* of programming, which characterizes programming languages like Lisp and Haskell, and logical systems like Church's λ-calculus.[22] As a matter of fact, the currying and execution of 12+81 could be naturally expressed in λ-calculus by the following β-reductions, i.e., computing steps:

$$\begin{aligned}(\lambda x \lambda y \cdot y + x)(81)(12) &\leadsto \boxed{(\lambda y \cdot y + 81)}(12)\\ &\leadsto 12 + 81\\ &\leadsto 93.\end{aligned}$$

Thus, if one were to read formulae as programs, as claimed before, then

$$(\lambda x \cdot \lambda y \cdot y + x)(81)(12),$$

or equivalently

$$(x \mapsto (y \mapsto y + x))(81)(12),$$

would be a better candidate than the more misleading notation $12 + 81$, which does not fully reveal the asymmetry of operands during the execution of the algorithm.

The concept of functional style can thus be used to describe algorithms—hence the computing tools which use them as well.[23]

[21](Backus, 1978).

[22]See (Church, 1932, p. 352): "Adopting a device due to Schönfinkel [1924], we treat a function of two variables as a function of one variable whose values are functions of one variable, and a function of three of more variables similarly."

[23]Even in the case of two numbers alternatively acting on each other, as for the greatest common divisor (Fig. 6), the computing style of the abacus remains functional. Indeed, no other value than the input data is created: only the input data are used to transform each other. There is in this algorithm a kind of parallelism, since two processes are executed simultaneously, interacting with each other; but this parallelism is no less *functional*, for every number is alternatively an act or an object for the other.

2 and 2 make 4				
2 and 3 make 5	3 and 3 make 6			
2 and 4 make 6	3 and 4 make 7	4 and 4 make 8		
2 and 5 make 7	3 and 5 make 8	4 and 5 make 9	⋱	
…	…	…	…	⋱
2 and 9 make 11	3 and 9 make 12	4 and 9 make 13	…	…

Fig. 11 Fibonacci's addition table

4 Pedagogical Strategies and Semantics of Programming Languages

The discrepancy thus observed between arithmetical and algorithmic properties may be found puzzling. Light could be shed on it by the distinction between two kinds of semantics—denotational or operational—for programming languages. Two programs are said to be *denotationally equivalent* if they produce the same result, and *operationally equivalent* if their respective executions follow the same steps.

This distinction will moreover highlight different strategies for teaching the use of computing tools. Some pedagogies indeed are result-oriented (or denotational), others are method-oriented (or operational).

Let us have a look at Fibonacci's addition algorithm in the *Liber Abaci*.[24] An addition table is displayed, which is, so to speak, "triangular:" its height decreases from column to column (Fig. 11). On this table, "2 and 4" is easy to find, but "4 and 2" is missing. One actually does not need to write the latter, when computing an addition on paper, since the only thing we want to know is the result of those elementary additions (from $1 + 1$ to $9 + 9$); as addition is commutative, there is no need to write twice the same result, and the addition table needs not be square. The diagonal of the addition table is an axis of symmetry. Being a table of pure results, Fibonacci's addition table is denotational: it does not show how to (operationally) compute those additions, but only what the results are (denotationally). This is why the addition table (or half the table) has to be learned by heart.

By contrast, the addition method which is described in Cheng Dawei's *Suanfa Tongzong* under the name of *Jiujiu bashiyi* (九九八十一)[25] does not *state* as explicitly as it *shows* that $4 + 2 = 6$ and that $2 + 4 = 6$; it just describes how to get the result when, having 2 on a given rod, one wants to add 4, or when one

[24](Fibonacci, 2002, Chapter I, p. 21).

[25](Cheng Dawei 程大位, 1993, p. 1231–1232). *Jiujiu bashiyi* means "9 times 9 makes 81;" the exercise actually consists of 81 successive operations. The same exercise can be found, under different names, in other treatises on the abacus: (Xu Xinlu 徐心魯, 1993, p. 1143–1147); (Ke Shangqian 柯尚遷, 1993, p. 1171–1174); (Huang Longyin 黃龍吟, 1993, p. 1428–1429). Chen Yifu wrote a detailed analysis of those texts in (Chen Yifu, 2013, Part I).

Chinese text	Instructions	Operation
二下五除三 ("two down five remove three")	[In order to add] two, [move] down [one] quinary [bead] [and] remove three [unary beads]	$[4] + 2 = [4] + 5 - 3$
四下五除一 ("four down five remove one")	[In order to add] four, [move] down [one] quinary [bead] [and] remove one [unary bead]	$[2] + 4 = [2] + 5 - 1$

Fig. 12 Different operations for $4 + 2$ and $2 + 4$

Chinese text	Instructions	Operation
四下五除一 ("four down five remove one")	[In order to add] four [move] down [one] quinary [bead] [and] remove one [unary bead]	$[2] + 4 = [2] + 5 - 1$
四下五除一 ("four down five remove one")	[In order to add] four [move] down [one] quinary [bead] [and] remove one [unary bead]	$[3] + 4 = [3] + 5 - 1$

Fig. 13 The same operation for $2 + 4$ and $3 + 4$

has 4 and wants to add 2. What is surprising is that, even though $4 + 2$ and $2 + 4$ are, of course, denotationally equivalent (i.e., they give the same result), there are two *operationally different* ways of computing this sum, depending on the order of the operands, as can be seen on Fig. 12. If those instructions were to be laid out as a square table, the diagonal would not be an axis of symmetry.

Even more surprisingly, even though $2 + 4$ and $3 + 4$ are denotationally different (they give different results, respectively 6 and 7), they are *operationally equivalent* on the abacus, for the manual operations and the instructions are the same, as can be seen on Fig. 13.

Thus, two operations of the abacus can be denotationally equivalent and operationally different (as $4 + 2$ and $2 + 4$), or denotationally different and operationally equivalent (as $2+4$ and $3+4$). In all cases, Cheng Dawei's instructions in the *Jiujiu bashiyi* follow the operational viewpoint. The pedagogical style of this exercise for the abacus is therefore not denotational, but operational.

Concepts taken from semantics of programming languages can thus be used to describe, if not how practitioners represent themselves what computation is, at least how they teach it.

5 Conclusion

Computer science is much more than the science of computers. Some of its concepts, such as programming paradigms, denotational and operational semantics, variable assignment, data structures and currying, can be used as historical tools.

These concepts allow us to describe how various computing tools are really used in practice, and how they are taught.

Such a method can be extended to other computing tools, such as the Greek abacus, the logarithm table, the slide rule, etc. A systematical inquiry into the representability of arithmetical properties on computing tools would lead, if not to a general classification—as long as we have no natural classification of programming languages—at least to some local comparison between existing computing tools.

If computer science and computability theory were to find such applications in the most concrete history of mathematics, their vocabulary would fruitfully integrate the toolbox of historians and philosophers of mathematics, especially of those interested in mathematical practices.

Acknowledgements The author thanks Karine Chemla, Baptiste Debrabant, Denis Roegel, the editors and the anonymous reviewers for their benevolent help.

References

Backus, J. (1978). Can programming be liberated from the von Neumann style?: A functional style and its algebra of programs. *Communications of the ACM, 21*(8), 613–641.

Chemla, K. (1996). Positions et changements en mathématiques à partir de textes chinois des dynasties han à song-yuan. quelques remarques. *Extrême-Orient, Extrême-Occident, 18*, 115–147.

Chemla, K., & Shuchun, G. (Eds.). (2004). *Les Neuf Chapitres. Le Classique mathématique de la Chine ancienne et ses commentaires*. Paris: Dunod.

Chen, Y. (2013). *L'Étude des différents modes de déplacement des boules du boulier et de l'invention de la méthode de multiplication* Kongpan Qianchengfa *et son lien avec le calcul mental* (PhD thesis). Université Paris-Diderot, Paris.

Cheng Dawei 程大位 (1993). Suanfa tongzong 算法統宗. In Guo Shuchun 郭書春 (Ed.), *Zhongguo kexue jishu dianji tonghui. Shuxue juan* 中國科學技術典籍通彙. 數學卷 (Vol. 2, pp. 1213–1421). Zhengzhou. Henan jiaoyu chubanshe 河南教育出版社.

Church, A. (1932). A set of postulates for the foundation of logic part i. *Annals of Mathematics, 33*(2), 346–366.

Fibonacci, L. (2002). *Fibonacci's Liber abaci : A translation into modern English of Leonardo Pisano's Book of calculation*. Springer, New York.

Huang Longyin 黃龍吟 (1993). Suanfa zhinan 算法指南. In 郭書春, G. S. (Ed.), *Zhongguo kexue jishu dianji tonghui. Shuxue juan* 中國科學技術典籍通彙. 數學卷 (Vol. 2, pp. 1423–1453). Zhengzhou. Henan jiaoyu chubanshe 河南教育出版社.

Ke Shangqian 柯尚遷 (1993). Shuxue tonggui 數學通軌. In Guo Shuchun 郭書春 (Ed.), *Zhongguo kexue jishu dianji tonghui. Shuxue juan* 中國科學技術典籍通彙. 數學卷 (Vol. 2, pp. 1165–1212). Zhengzhou. Henan jiaoyu chubanshe 河南教育出版社.

Knuth, D. E. (1997). *The art of computer programming. Vol. 1 : Fundamental algorithms*. Reading, MA: Addison-Wesley.

Rechenberg, P. (1990). Programming languages as thought models. *Structured Programming, 11*, 105–115.

Schärlig, A. (2001). *Compter avec des cailloux. Le calcul élémentaire sur l'abaque chez les anciens Grecs*. Lausanne: Presses polytechniques et universitaires romanes.

Schönfinkel, M. (1924). Über die Bausteine der mathematischen Logik. *Mathematische Annalen, 92*, 305–316.

Sunzi 孫子 (1993). Sunzi suanjing 孫子算經. In Guo Shuchun 郭書春 (Ed.), *Zhongguo kexue jishu dianji tonghui. Shuxue juan* 中國科學技術典籍通彙. 數學卷 (Vol. 2, pp. 225–245). Zhengzhou. Henan jiaoyu chubanshe 河南教育出版社.

Turing, A. M. (1936). On computable numbers, with an application to the *Entscheidungsproblem*. *Proceedings of the London Mathematical Society, 2*(42) 230–265.

von Neumann, J. (1993). First draft of a report on the EDVAC. *IEEE Annals of the History of Computing, 15*(4), 27–75.

Xu Xinlu 徐心魯 (1993). Panzhu suanfa 盤珠算法. In Guo Shuchun 郭書春 (Ed.), *Zhongguo kexue jishu dianji tonghui. Shuxue juan* 中國科學技術典籍通彙. 數學卷 (Vol. 2, pp. 1141–1164). Zhengzhou. Henan jiaoyu chubanshe 河南教育出版社.

Mathematics in Engineering and Science

Stig Andur Pedersen

Abstract

Mathematics is an important aspect of natural science and engineering and many new mathematical concepts and theories have come about when researchers have been formulating and solving scientific or engineering problems. This interaction between mathematics and science/engineering took a new form towards the end of twentieth century in connection with the introduction of digital computers. In fact, some scientists believe that a new form of doing science has appeared: computational science. Norman J. Zabusky, e.g., argues that we are in the midst of a computational revolution that will change science and society as dramatically as the agricultural and industrial revolutions did (Zabusky, Phys Today 40(10), 1987). We shall discuss in what sense it is reasonable to talk about a new style of scientific reasoning and what this will mean for mathematical practice.

Mathematics is a deductive science, and mathematical results are presented in deductive form. As a rule, you are able to follow the argumentation in a mathematical paper from basic axioms to the conclusion, but this does not mean that practical mathematical reasoning follows the same deductive pattern. In real mathematical reasoning, you will, of course, find all kinds of deviations from basic assumptions. The deductive form is the important final presentation of a valid mathematical argument, but it does not reflect the creative process of the mathematician. This paper will begin with a characterisation of practical mathematical reasoning. It will continue with some historical examples from applied mathematics that have

---

S.A. Pedersen (✉)
Section for Philosophy and Science Studies, Roskilde University, PO Box 260, DK-4000 Roskilde, Denmark
e-mail: sap@ruc.dk

S. Ju et al. (eds.), *Cultures of Mathematics and Logic*, Trends in the History of Science, DOI 10.1007/978-3-319-31502-7_4

played an important role in the development of scientific computation. Finally, it will argue that a new style of doing science—computational science, not to be confused with computer science—has appeared as a fusion of numerical analysis, applied mathematics and computer science.

1 The Nature of Mathematical Reasoning

Richard P. Feynman maintains that there are two different ways of doing mathematics: the Babylonian tradition and the Greek or Euclidian tradition, and he emphasizes that physicists need the Babylonian one.

The Babylonian style is characterised in the following way:

> In Babylonian schools in mathematics the student would learn something by doing a large number of examples until he caught on to the general rule. Also he would know a large amount of geometry, a lot of the properties of circles, the theorem of Pythagoras, formulae for the areas of cubes and triangles; in addition, some degree of argument was available to go from one thing to another. (Feynman, 1992, p. 46)

and the Euclidian one in this way:

> But Euclid discovered that there was a way in which all the theorems of geometry could be ordered from a set of axioms that were particularly simple. ... The most modern mathematics concentrates on axioms and demonstrations within a very definite framework of conventions of what is acceptable and what is not acceptable as axioms. (Feynman, 1992, p. 46)

Arguing in favour of the Babylonian style, Feynman refers to the gravitational field and asks which is the more basic: the fact that "the force is towards the sun, or ··· that equal areas are swept out in equal times" (Feynman, 1992, p. 47). From one point of view, the force statement is better, Feynman says, because it is possible to generalise that statement. If we consider a system of many particles, it is still possible to make a strong case for the force statement but the orbits are no longer ellipses. Therefore, it makes sense to have the force law as an axiom whereas Kepler's law is too specific and must be generalised in order to hold true for more general systems. Kepler's law is a special case of conservation of angular momentum, and conservation of angular momentum holds quite generally.

The law of conservation of angular momentum is interesting because it can be derived from a specific case and then extended to virtually all fields of physics, and, as Feynman says,

> We have these wide principles which sweep across the different laws, and if we tie the derivation too seriously, and feel that one is only valid because another is valid, then we cannot understand the interconnections of the different branches of physics. (Feynman, 1992, p. 49)

Only if we reach a point where physics is complete and we know all the laws, may we be able to start with definite axioms and deduce the rest. However, that is an unreachable ideal state of affairs and "...while we do not know all the laws, we can use some to make guesses at theorems which extend beyond the proof". (Feynman, 1992, p. 50)

The situation is similar in mathematics. For instance, we can axiomatise Euclidian geometry by using Hilbert style axioms, that is, incident and congruence axioms etc. and from these axioms deduce the existence of the three-dimensional symmetry group. However, invariance with respect to a group is a general phenomenon which holds in many other cases and the prospect of defining and studying classes of geometries based on the concept of transformation group leads to many new possibilities as demonstrated by Felix Klein's Erlanger Program. If we consider geometry from the viewpoint of transformation groups, it is possible to build on methods from algebra, group theory and metric spaces.[1]

Several researchers have observed similar "Babylonian features" in mathematics. In the well-known paper *The Unreasonable Effectiveness of Mathematics in the Natural Sciences* Eugene Wigner writes:

> ··· mathematics is the science of skillful operations with concepts and rules invented just for this purpose. The principal emphasis is on the invention of concepts. Mathematics would soon run out of interesting theorems if these had to be formulated in terms of the concepts which already appear in the axioms. (Wigner, 1967, p. 224)

So, the essence of mathematical reasoning is not deductions from given axioms. It consists of creating new concepts and rules which may lead to new insights. Some of these new concepts may then appear in many unexpected situations. For instance, the idea of group action is important in many areas of mathematics far removed from geometry. It is, in fact, a main point in Wigner's paper to show how local ideas in mathematics often show up in quite unexpected surroundings, and the strong interplay between different areas of the subject may even be considered a characteristic feature of modern mathematics. Think, e.g., of the applications of functional analysis in number theory or the role of Riemann's ζ-function in many different areas of mathematics.[2]

[1]Cf., e.g., the elegant presentation by Rees (1988). Rees's presentation is Babylonian:

> ··· I have taken a concrete viewpoint rather than an axiomatic one. The view that I take is that mathematical objects exist and should be studied, they are not arbitrarily defined as the axiomatic approach might suggest. This is the view of the vast majority of mathematicians in their own work and it is a pity that this does not come across in more undergraduate courses. (Rees, 1988, p. vi)

[2]Sir Michael Atiyah characterises mathematics in the twentieth century in this way: The first half of this century is called "the 'era of specialisation', the era in which Hilbert's approach, of trying to formalise things and define them carefully and then follow through on what you can do in each field, was very influential" (Atiyah, 2002, p. 13). The second half of the twentieth century is called

If it were impossible to introduce new concepts that reached far beyond the scope defined by the concepts introduced by the axioms, it would, according to Wigner, be impossible to formulate interesting new theorems (Wigner, 1967). Besides, as Hamming argues in a paper inspired by Wigner, there are many results in mathematics that are independent of the axioms:

> If the Pythagorean theorem were found to not follow from the postulates, we would again search for a way to alter the postulates until it was true. Euclid's postulates came from the Pythagorean theorem, not the other way. For over thirty years I have been making the remark that if you came into my office and showed me a proof that Cauchy's theorem was false I would be very interested, but I believe that in the final analysis we would alter the assumptions until the theorem was true. (Hammig, 1980, p. 87)

So, new mathematical theorems are not latently lying in the axioms ready to be deduced. They are, in a sense, constructed by introducing new concepts and ideas. In many cases, we are inspired to develop new concepts by trying to describe regularities in nature. Wigner describe this creative process by formulating what he calls *the empirical law of epistemology*:

> ... the "laws of nature" being of almost fantastic accuracy but of strictly limited scope. I propose to refer to the observation which these examples illustrate as the empirical law of epistemology. Together with the laws of invariance of physical theories, it is an indispensable foundation of these theories. Without the laws of invariance the physical theories could have been given no foundation of fact; if the empirical law of epistemology were not correct, we would lack the encouragement and reassurance which are emotional necessities, without which the "laws of nature" could not have been successfully explored. (Wigner, 1967, p. 233)

According to Wigner's empirical law of epistemology, it is possible to formulate laws of nature by using mathematics which will hold with "almost fantastic accuracy but [are] of strictly limited scope". It is the empirical law of epistemology that creates the sense of encouragement and reassurance which is a necessary condition for our successful exploration of nature.

As a physicist Wigner was primarily interested in mathematical descriptions of nature. However, much modern mathematics is developed internally without any reference to physics. It has its roots not in descriptions of natural invariances, but in internal structures within mathematics itself. Set theory is a case in point. Cantor's definition of the ordinal numbers was a consequence of his transfinite iteration of the process of taking limit points of a point set of the real line.[3] In fact, Cantor saw a fruitful way of extending inductive definitions into the transfinite, and he was able to isolate and study this and other set theoretical features. This is, in a way, similar to how Kepler's law could be seen as a special case of angular

"the 'era of unication', where borders are crossed over, techniques have been moved from one field into the other, and things have become hybridised to an enormous extent" (Atiyah, 2002, p. 14).

[3]Cf. *Über unendliche linear Punktmannigfaltigkeiten* in (Cantor, 1962).

momentum conservation. Cantor's early work on set theory is a good illustration of how mathematics develops by identifying new patterns and introducing new concepts that describe these patterns. The final result, in Cantor's case the ordinal numbers, is the result of a long process consisting of conjectures, problems, ideas, etc. V.I. Arnold illustrates this phenomenon with a wonderful metaphor:

> When you are collecting mushrooms, you only see the mushroom itself. But if you are a mycologist, you know that the real mushroom is in the earth. There's an enormous thing down there, and you just see the fruit, the body that you eat. In mathematics, the upper part of the mushroom corresponds to theorems that you see. But you don't see the things which are below, namely *problems, conjecture, mistakes, ideas, and so on*. (Arnold, 2006, p. 19)

It is difficult to identify the driving forces in the development of mathematics, especially when we bear in mind that mathematical praxis is more Babylonian than Euclidian and that theorems and deductive structures are only a minor part of the activity. However, one central driving force is the search for laws and regularities independently of whether these laws and regularities come from physics, mathematics itself or from other sources. It is therefore a bad idea to differentiate radically between pure and applied mathematics, a point made very clearly by Mark Kac and Stanisław M. Ulam:

> It should perhaps be stressed again that the boundaries between mathematics and the many disciplines to which it is applied are seldom sharply drawn. Nothing but impoverishment can be expected from the unfortunately rather frequent current efforts to isolate a body of 'pure' mathematics from the rest of scientific endeavour and to let it feed only on itself. (Kac and Ulam, 1979, p. 180)

It should now be clear that practical mathematical reasoning is mainly Babylonian in style whereas the Euclidian style is more dominant when mathematical results are presented in theoretical papers. We can summarise our view as follows:

1. A final mathematical proof must comply with classical (or intuitionistic) logic. It is, of course, a fundamental criterion of validity that a mathematical proof is logically valid, and it is from the proof that it is possible to see the deeper relationship between the theorem and fundamental mathematical principles. Thus, logical correctness (consistency) is an essential aspect of mathematical truth. Without a logical correct proof, a statement remains a hypothesis and not a theorem.
2. Proof constructions often require the introduction of new concepts. David Hilbert, e.g., said that a characteristic part of mathematical reasoning was the introduction of ideal elements. Typical examples are the introduction of the complex unit, i, and the introduction of ordinal numbers. Modern examples are also abundant, such as K-groups, distributions, noncommutative geometries, etc.
3. New important theorems, concepts, and hypotheses often codify and generalise problems appearing in science and engineering or internally in mathematics itself. For example, the method of finite elements for numerical solutions of

partial differential equations was invented by engineers and then taken up by numerical analysts; Noether's theorem on the relationship between conservation laws and symmetries was inspired by issues relating to the mathematical formulation of energy conservation in general relativity theory; the concepts of category and functor were introduced in algebraic topology in order to cope with the interplay between algebraic and topological structures.

2 Some Historical Examples

The interaction between mathematics, physics and engineering has always been strong. Physics and engineering have always been an important source of new mathematical ideas, and new mathematical theories and results often find their way into physical and engineering applications. As Wigner expresses it, "mathematical concepts turn up in entirely unexpected connections" and "they often permit an unexpectedly close and accurate description of the phenomena in these connections" (Wigner, 1967). However, there is a perennial problem involved in the application of mathematics, namely, that it is usually impossible to solve analytically the equations that describe the phenomena being studied. In nearly all applications, we need to settle for approximative solutions, and in many cases our computational capabilities are too limited. This condition of applied mathematics was felt strongly in hydrodynamics at the beginning of the last century and was eloquently described by Theodore von Kármán in his Gibbs lecture from 1939:

> Due to this failure of the method we do not get an answer for one of the fundamental questions of the hydrodynamics of real fluids, that is : What is the flow pattern of a real fluid around a submerged body in the limiting case $\nu \rightarrow 0$? As a matter of fact this problem is still not solved. Consider, for example, two-dimensional flow around a circular cylinder. We are not able to decide whether the flow pattern for $\nu \rightarrow 0$ approaches the potential flow of a nonviscous fluid or a stationary flow pattern consisting of a vortex-free region and a wake with continuously distributed vorticity, as suggested by Oseen, or a nonstationary flow pattern with concentrated vortex columns of alternating circulation, a flow pattern treated by the present author. It seems that we have here an example in which the analytical methods are not sufficient, at least at the present time, to solve a problem of purely analytical character. (von Kármán, 1940, p. 664)

Kármán characterises the situation in hydrodynamics during the time just before the first electronic computers appeared, and, as we shall see, the appearance of electronic computers would drastically change the condition of applied mathematics.

The understanding that a computer device is able to store and operate on its own programs appeared in the mid 1930s as the result of the efforts of Kurt Gödel (1931) and Alan Turing (1937). By introducing Gödel numbers, Gödel made it clear that a theory or axiom system could be reflexive and therefore able to model itself. By a similar construction, Turing introduced the universal Turing machine.

At the beginning the idea of a computational system or the concept of what we should understand by an algorithm was mainly motivated by philosophical and mathematical issues of what in principle could be calculated or reached by constructive methods. Since the end of the nineteenth century mathematicians have formulated theorems which could be proved by nonconstructive methods—e.g., by using the axiom of choice or by giving indirect proofs—but where no constructive proofs were known. For example, one of David Hilbert's famous problems from the International Conference of Mathematicians in 1900 was whether or not it would be possible to find a general algorithm which when given a finite set of Diophantine equations would either give a solution or say that the system did not have a solution. The precise definition of algorithm first appeared in the mid 1930s through the research by Gödel, Turing and many others.[4]

In addition to the theoretical development in the late 1930s, there was, as shown by von Kármán's Gibbs lecture, a definite need for computational methods . This need intensified further with the advent of World War II. The birth of the first electronic computers was, in a sense, the result of a fortuitous alliance between highly theoretical and philosophical advances and the practical needs of especially military research during and after Wold War II.

There were many researchers involved in the development of the electronic computer. However, John von Neumann stands out particularly as he both contributed to the theoretical design of the computer and also set an example for using the computer for scientific calculations.[5] Originally von Neumann was a theoretical mathematician doing work related to Hilbert's school in Göttingen. However, his association with *The Institute of Advanced Studies* at Princeton University and his acquaintance with military research changed his focus. Von Neumann describes himself as follows:

> It was through him [Robert Kent, a senior official at the Ballistics Research Laboratory, Maryland] that I was introduced to applied science. Before this I was, apart from some lesser infidelities, essentially a pure mathematician, or at least a very pure theoretician. Whatever else may have happened in the meantime, I have certainly succeeded in losing my purity.[6]

Besides being involved in the design of computers, he became a leading figure in the development of numerical methods dedicated to applications on new fast computers.

We know that Turing and von Neumann met during the mid 1930s and that von Neumann thought very highly of Turing's work[7] but it is not clear whether von Neumann got the idea of the program-storing computer from Turing. It was von Neumann's description of *the von Neumann architecture* in *First Draft of a Report on the EDVAC* dated 30 June 1945 that became the leading design idea for the modern electronic computer. However, it is probably more correct to say that von

[4]For instance, Alonso Church, Steve Kleene, Emil Post, Andrey Markov and many others.

[5]A study of John von Neumann's role in this development can be found in Aspray (1990).

[6]Quoted from Aspray (1990, p. 26).

[7]Cf. (Aspray, 1990, pp. 176–178).

Neumann's report was based on joint ideas developed by the staff at the Moore School. For instance, J. Presper Eckert and John W. Mauchly wrote

> Dr. von Neumann has also written a preliminary report in which most of the results of earlier discussions are summarized. In his report, the physical structure and devices *proposed by Eckert and Mauchly* are replaced by idealized elements to avoid raising engineering problems which might distract attention from the logical considerations under discussion.[8]

In any event, through his strong scientific involvement at the Institute of Advanced Studies and his practical involvement in model-building activities of military relevance, von Neumann came to set the scene for the development of both the computer architecture and the application of computers for scientific purposes.

One distinctive area where the new electronic computer came to show its numerical merits was meteorology. It was a widely held view in the 1940s that the partial differential equations derived in meteorology were far beyond the reach of numerical methods. The equations are non-linear and involve several scales. Too many iterations on different scales would be required and it would be difficult to control errors. One of the first successes of computer-based numerical calculations was outlined in the paper by J.G. Charney, R. Fjörtoft and J. von Neumann on the numerical integration of the barotropic vorticity equation from 1950 (Charney et al., 1950). It is characteristic that this paper only handled a rather simplified aspect of the weather forecast problem, namely the phenomenon of vorticity. The equations treated were the following:

$$\frac{\partial \eta}{\partial t} + \bar{v} \cdot \nabla \eta = 0$$

$$\eta = \zeta + f = \text{absolute vorticity}$$

$$\zeta = \text{vertical component of the curl of } \bar{v}$$

$$f = 2\Omega \sin \phi = \text{Coriolis parameter}$$

where $\bar{v}$ was the wind velocity, Ω the angular frequency of the Earth's rotation, and ϕ the latitude. The paper was characterised by the authors themselves in the following way:

> A method is given for the numerical solution of the barotropic vorticity equation over a limited area of the earth's surface. The lack of a natural boundary calls for an investigation of the appropriate boundary conditions. These are determined by a heuristic argument and are shown to be sufficient in a special case. Approximate conditions necessary to insure the mathematical stability of the difference equation are derived. (Charney et al., 1950, p. 237)

A typical finite difference grid used in the computation is depicted in Fig. 1.

[8]Quoted from Aspray (1990, p. 42).

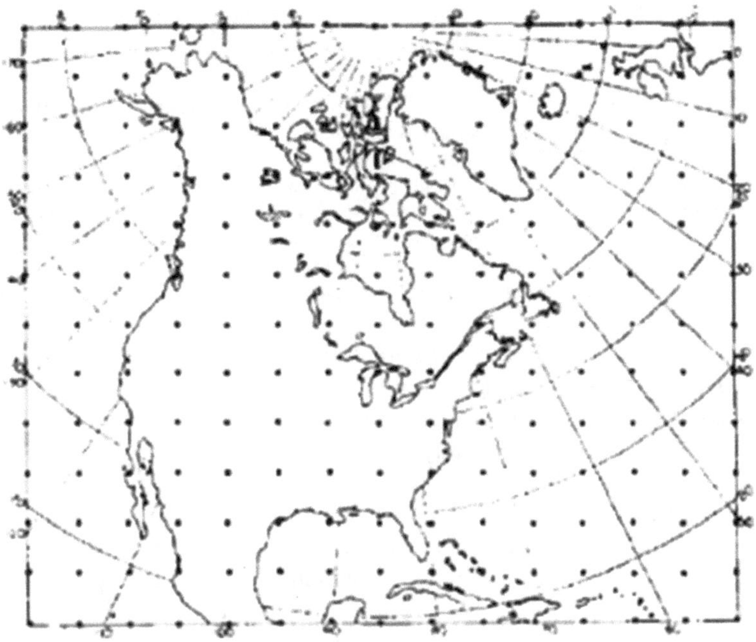

Fig. 1 Finite difference grid used in the computation

The calculations of the paper were made on the ENIAC computer. They were to some extent a success. The first forecast for 5 January 1950 was not satisfactory:

> The forecast of January 5, in which the principal system was an intense cyclone over the United States, was uniformly poor. The forecast gave much too small a displacement of the cyclone and also distorted its shape, and the predictions of the other motions were equally inaccurate. (Charney et al., 1950, p. 245)

But the next one was much better:

> On the other hand, the January 30 forecast contained a number of good features. The displacement and amplification of the trough over the United States at about 1100 W was well predicted, as was the large scale shifting of the wind from NW to WSW and the increase in pressure over eastern Canada. The displacement of the axis of the major trough over the eastern United States and Canada was correctly predicted, but the strong circulation that developed at its southern extremity was not. Proceeding eastwards we find that the amplification of the trough over the North Sea together with the characteristic breakthrough of the northwesterly winds and the corresponding destruction over France of the eastern nose of the anticyclone was predicted approximately. This is shown by the agreement of the predicted with the observed height changes over western Europe. (Charney et al., 1950, p. 276)

This first successful numerically calculated weather prediction has been redone on modern computers. During his Starr Lecture in the 1980s, George Platzman arranged with IBM to repeat one of the ENIAC forecasts. The result was commented upon by Peter Lynch in the following way:

> The algorithm of CFvN [Charney, Fjörtoft, and von Neumann]was coded on an IBM 5110, a desktop machine then called a portable computer or "PC" (having a tiny fraction of the power of a modern PC). The program execution was completed within the hour or so of Platzman's lecture. This implies a 24-fold speedup over the best rate achievable for ENIAC. The program eniac.m was run on a Sony Vaio (model VGN-TX2XP) with MATLAB version 6. The main loop of the 24 h forecast ran in about 30 ms. One may question the precise significance of the time ratio—about three million to one—but it certainly indicates the dramatic increase in computing power over the past half-century. (Lynch, 2008, pp. 50–51)

The paper by Charney, Fjörtoft and von Neumann raised the hope that electronic computers might make numerical weather forecasts possible,[9] and as the extract suggests this hope has increased over the years. These days, computational meteorology is an essential part of the daily weather forecast.

As indicated by von Kármán in his Gibbs lecture, hydrodynamics desperately needed numerical methods in the 1930s. It still does but the work by Charney, Fjörtoft and von Neumann together with many others started a new development based on computer simulation, and, as the quotation from Lynch's paper illustrates, current numerical possibilities have increased so radically that it makes sense to talk about a paradigm shift in applied mathematics.

It was not only hydrodynamics that benefitted from this radical shift. Many other fields of applied mathematics experienced the same shift. We shall discuss another example where experiments with computational methods led to new mathematical discoveries.

In the years after World War II, the Italian physicist Enrico Fermi made frequent summer visits to Los Alamos. He became interested in the development and potentialities of electronic computing machines and held many discussions with the mathematician Stanisław M. Ulam on the kind of future problems which could be studied using such machines.[10] They decided to try a variety of problems for

---

[9]Peter Lynch describes the results in the following way

> The results were sufficiently encouraging that numerical weather prediction became an operational reality within about five years. (Lynch, 2008, p. 45)

In a lecture to the National Academy of Science in 1955 Charnet said

> The advent of the large-scale electronic computer has given a profound stimulus to the science of meteorology. For the first time the meteorologist possesses a mathematical apparatus capable of dealing with the large number of parameters required for determining the state of the atmosphere and of solving the nonlinear equations governing its motion. (Charney, 1955, p. 798)

[10]Cf. Stanisław Ulam's introduction to the paper *Studies of non linear problems* by Fermi, Pasta, and Ulam, (Segré, 1965, p. 977).

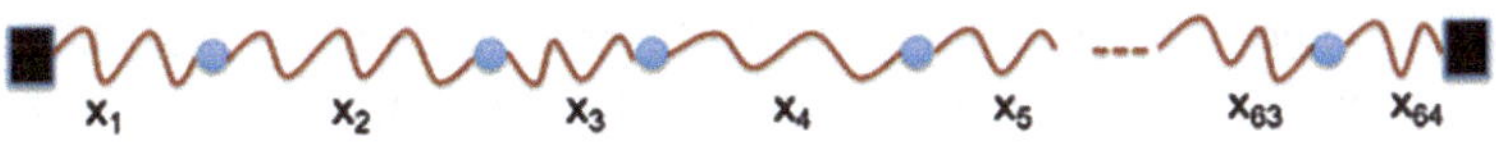

Fig. 2 Non-linear system studied by Fermi, Pasta and Ulam

heuristic work where, in the absence of closed analytic solutions, experimental work on a computing machine would perhaps contribute to the understanding of properties of solutions:

> This could be particularly fruitful for problems involving the asymptotic long time or "in the large" behavior of non-linear physical systems. In addition, such experiments on computing machines would have at least the virtue of having the postulates clearly stated. This is not always the case in an actual physical object or model where all the assumptions are not perhaps explicitly recognized.
> Fermi expressed often a belief that future fundamental theories in physics may involve non-linear operators and equations, and that it would be useful to attempt practice in the mathematics needed for the understanding of non-linear systems.[11]

One of the first systems they studied was a system of 64 mass points connected with non-linear strings as shown in Fig. 2. The results were reported in the joint paper by Fermi, Pasta and Ulam entitled *Studies of non linear problems*.[12]

The system is described by one of the equations:

$$\ddot{x}_i = (x_{i+1} + x_{i-1} + 2x_i) + \alpha[(x_{i+1} - x_i)^2 + (x_i - x_{i-1})^2]$$
$$\ddot{x}_i = (x_{i+1} + x_{i-1} + 2x_i) + \beta[(x_{i+1} - x_i)^3 + (x_i - x_{i-1})^3]$$

where $i = 1, 2, \ldots, 64$, x_i denotes the displacement of the i-th point from its original position, α denotes the coefficient of the quadratic term in the force between the neighbouring mass points and β that of the cubic term. (Segré, 1965).

The first equation defines a second order non-linearity and the second equation a third order non-linearity. The first equation leads to the following expression for the sum of kinetic and potential energies

$$E_{x_i}^{\text{kin}} + E_{x_i}^{\text{pot}} = \frac{1}{2}\dot{x}_i^2 + \frac{(x_{i+1} - x_i)^2 - (x_i - x_{i-1})^2}{2}$$

where contributions to potential energy from quadratic or higher terms in the force are neglected.

The purpose of the study was to see how the string when started in a simple configuration would "assume more and more complicated shapes, and, for t tending to infinity, would get into states where all the Fourier modes acquire increasing

---

[11]Stanisław Ulam's introduction to the paper *Studies of non linear problems* by Fermi, Pasta and Ulam, (Segré, 1965, p. 977).

[12]The paper was written in 1954 but first published in (Segré, 1965).

importance" (Fermi et al., 1965, p. 980). In order to see this, Fermi, Pasta and Ulam performed a Lagrangian change of variables $(x_i, \dot{x}_i) \rightarrow (a_k, \dot{a}_k)$, $i = 1, 2, \ldots, 64; k = 1, 2, \ldots, 64$, by using the formula

$$a_k = \sum_{i=1}^{64} x_i \sin \frac{ik\pi}{64}.$$

With this transformation we get the following representation of the total energy with respect to the Fourier modes

$$E_{a_k}^{\text{kin}} + E_{a_k}^{\text{pot}} = \frac{1}{2}\dot{a}_k{}^2 + 2a_k^2 \sin^2 \frac{\pi k}{128}.$$

These equations were solved on the MANIAC machine at Los Alamos and the results surprised Fermi:

> The results of the calculations (performed on the old MANIAC machine) were interesting and quite surprising to Fermi. He expressed to me the opinion that they really constituted a little discovery in providing intimations that the prevalent beliefs in the universality of "mixing and thermalization" in non-linear systems may not be always justified.[13]

Figure 3 is a graphical representation made by Ulam and Fermi of the energies in the first five modes of the system. The figure is based on 30,000 computation cycles. The initial form of the string was a single sine wave.

Fermi, Pasta and Ulam comment on the results in the following way:

> Instead of a gradual, continuous flow of energy from the first mode to the higher modes, all of the problems show an entirely different behavior. Starting in one problem with a quadratic force and a pure sine wave as the initial position of the string, we indeed observe initially a gradual increase of energy in the higher modes as predicted (e.g., by Rayleigh in an infinitesimal analysis). Mode 2 starts increasing first, followed by mode 3, and so on. Later on, however, this gradual sharing of energy among successive modes ceases. Instead, it is one or the other mode that predominates. For example, mode 2 decides, as it were, to increase rather rapidly at the cost of all other modes and becomes predominant. At one time, it has more energy than all the others put together! Then mode 3 undertakes this role. It is only the first few modes which exchange energy among themselves and they do this in a rather regular fashion. Fermi et al. (1965, p. 981)

However, this is not the end of the story. At a later time, mode 1 comes back very close to its initial value and the system seems to be almost periodic. This seemed to be a common feature of all the systems studied. Fermi expected that the systems would "thermalise" in the sense that the energy levels would level out. But that did

[13]Stanisław Ulam's introduction to (Fermi et al., 1965), (Segré, 1965, p. 977).

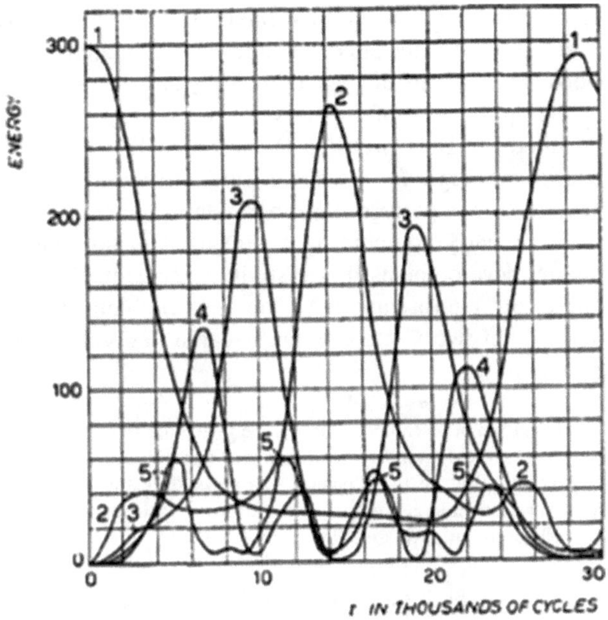

Fig. 3 Energies in the first five modes. This is the original figure from (Segré, 1965)

not happen:

> Instead of gradual increase of all the higher modes, the energy is exchanged, essentially, among only a certain few. It is, therefore, very hard to observe the rate of "thermalization" or mixing in our problem, and this was the initial purpose of the calculation. (Fermi et al., 1965, p. 981)

In 1961, the problem was recalculated on more modern and relatively faster computers for still longer periods of time. Jim Tuck and Mary Menzel found that the system also showed very long periods on top of the smaller ones.

> The total energy is concentrated again essentially in the first Fourier mode, but the remaining one or two percent of the total energy is in higher modes. If one continues the calculation, at the end of the next great cycle the error (deviation from the original initial condition) is greater and amounts to perhaps three percent. Continuing again one finds the deviation increasing—after eight great cycles the deviation amounts to some eight percent; but from that time on an opposite development takes place! After eight more, i.e., sixteen great cycles altogether, the system gets very close—better than within one percent to the original state! This super-cycle constitutes another surprising property of our non-linear system.[14]

Fermi, Pasta and Ulam completed their paper in 1954, the year when Fermi died. Fermi did see and discuss most of the results of the paper, but he did not see the

[14]Stanisław Ulam's introduction to (Fermi et al., 1965), (Segré, 1965, p. 978).

final version which was filed internally at Los Alamos in 1955 as *Document LA-1940*. However, the filed version was circulated among mathematical physicists. Both the problem of lack of mixing and thermalisation as well as the problem of the appearance of super-cycles were very surprising. This group of problems is called the FPU problem (Weisset, 1997). The FPU problem became a major guiding principle for the development of the theory of dynamical systems and the new method of performing experiments on computers (Zabusky, 2005).

These early computer experiments with non-linear systems led to several important developments in the study of dynamical systems which each gave rise to interesting interpretations of the FPU problem. However, as Thomas P. Weissert pointed out two views eventually came to dominate the understanding of the FPU problem:

> One group of researchers came to believe that FPU was a clear case of Kolmogorov-Arnold-Moser (KAM) stability, while another saw FPU as an example of Kortweg-deVries (KdV) solitons. Both of these phenomena are significant developments from dynamical systems theory that emerged in the years following FPU, and they are both intimately related to the FPU problem. (Weisset, 1997, pp. 29–30)

Anyhow, both the KAM and the KdV developments were inspired by the new impetus that the development and exploitation of the new electronic computers gave to applied mathematics and mathematical physics. They both represent mathematical research areas that heavily depended on the application of modern computers and also led to new developments of numerical algorithms.

The development of modern computational methods has influenced our way of doing mathematics in many other ways. However, this may be the subject of another article, as our current focus is applied mathematics and numerical methods.[15] The two examples provided here illustrate two central points, namely, how new computational methods have influenced our understanding of non-linear systems and how it has been possible to cope with data intensive and complex systems as the circulation of the atmosphere.

3 Computational Science: Emergence of a New Reasoning Style

The examples in the last section illustrate how numerical analysis achieved new impetus as electronic computers were introduced. Numerical analysis has always been an important aspect of applied mathematics and nearly all great mathematicians were excellent numerical analysts—think, e.g., of Newton, Euler, Laplace, Gauss and Bessel. However, new numerical methods were quite often developed

[15] Cf. the unpublished paper *Implications of Experimental Mathematics for the Philosophy of Mathematics* by Jonathan Borwein (citeseerx.ist.psu.edu/viewdoc/summary?doi=10.1.130.5689) and the papers in *Erkenntnis*, Vol. 68 (3).

outside mathematics in disciplines where comprehensive numerical methods were important, e.g., in astronomy, geodesy and meteorology. Furthermore, numerical analysis was not considered a separate field of speciality in mathematics before the middle of the last century. In the years after World War II, numerical institutes and departments began to appear at universities around the world. Naturally, the main reason was that the role of mathematical model building in engineering and science increased dramatically during and after the war. The computer made it possible to cope with very realistic models in engineering and science and they required more sophisticated numerical methods.

> Computers used to get smaller and faster, but in recent decades the direction changed towards faster computing using parallel processing, cyber computing, and using dedicated rather than all-purpose processors. Different platforms will require dedicated algorithms, which will take full advantage of new computing architectures. As a recent example, we mention the success of using graphical processing units (GPUs) in running large scale simulations much faster than multicore systems. At the same time, the resulting increase in computing power will enable us to simulate more than just nonlinear PDEs at a given scale; it will enable us to model *hierarchies* of scales. (Tadmor, 2012, p. 544)

These new developments will make it possible to model the interplay between global circulation models and highly localised dynamics which are of central importance in numerical weather prediction:

> The main aspect in these approaches goes beyond the numerical solution of a given model: petaflop computational platforms will enable actual modeling across the hierarchy of discrete scales. These developments will enable, in the context of numerical weather prediction, for example, a multiscale simulation of the interplay between the global circulation numerical model and highly localized dynamics. We shall then get closer to realizing the full potential behind von Neumann's vision ⋯, where "the entire computing machine is merely one component of a greater whole, namely, of the unity formed by the computing machine, the mathematical problems that go with it, and the type of planning which is called by both." (Tadmor, 2012, p. 544)

What Eitan Tadmor talks about here is a new style of doing scientific research. Scientific and engineering calculations are no longer just numerical issues. Scientific and engineering aspects of computing form a complex symbiosis between computer architecture, the study of computer languages and algorithms, numerical analysis and scientific theory. In many fields, such as high energy physics, meteorology, astronomy and fluid dynamics, progress depends on close interdisciplinary cooperation between numerical analysis, computer science and the relevant scientific field.[16]

In order to better understand the impact of these changes, it make sense to apply Ian Hacking's idea of a *reasoning style*.[17] A reasoning style is similar to a Kuhnean paradigm but comprises more features of a scientific praxis. So, by *a scientific*

[16]Cf. Zabusky (1987).

[17]Cf. Hacking (1992b).

reasoning style[18] we understand a comprehensive transformation of the way in which science is being understood and performed. The transformation covers the following elements:

- The transformation concerns several scientific disciplines.
- New institutions are formed that epitomise the new directions.
- It leads to new social organisations of the scientific practice.
- It leads to fundamental ontological and epistemological changes: new types of objects, evidence, classifications, laws or modalities, and ways of expressing scientific facts.

As an example of a new reasoning style, Ian Hacking mentions the probabilistic reasoning style which emerged during the middle of the nineteenth century.[19] It came about as a consequence of the numerous statistical societies founded in the 1830s. Sam Schweber and Matthias Wächter describe it in the following way:

> The avalanche of numbers gave a different feel to the world: it had become quantified and numbers and statistics ruled it. It was Mr. Gradkin's world. Concomitantly, the previously dominant determinist Weltanschauung became replaced by a view of the world in which probability and chance played an ever increasing role. The result was the emergence of a new statistical style, constituted by a plethora of abstract statistical entities and governed by autonomous statistical laws, which are 'used not only to predict phenomena but also to explain [them]'. (Schweber and Wächter, 2000, p. 584)

This probabilistic revolution not only gave rise to a new indeterministic *Weltanschauung*. It also gave rise to new institutions where statistical information was collected, and to new ways of organising scientific research and expressing scientific facts. Hence, the statistical probabilistic development in the nineteenth century resulted in a new reasoning style which also required comprehensive numerical methods.

In a similar way, we may say that the development of digital computation capability has led to a new reasoning style which we call *computational science*.[20] The development of computational science has transformed all scientific disciplines that require processing of comprehensive data sets and also disciplines where complex non-linear models are essential. New institutions in the form of centres for scientific computation have emerged. Computer experiments and simulations have provided new ways of obtaining empirical information, and computer simulation has become a new way of identifying and solving problems in science and engineering. New kinds of objects and concepts have been introduced due to numerical and computational analyses of complex non-linear systems. This is true of both meteorology and the FPU problem. This development was described by

[18]This is a slight modification of Hacking's original definition.

[19]Cf. Hacking (1992a).

[20]Not to be mixed up with computer science.

Norman J. Zabusky back in 1987:

> We are in the midst of a computational revolution that will change science and society as dramatically as the agricultural and industrial revolutions did. The discipline of computational science is significantly affecting the way we do hard and soft science.
>
> ...
>
> Supercomputers with ultrafast, interactive visualization peripherals have come of age and provide a mode of working that is coequal with laboratory experiments and observations and with theory and analysis. We can now grapple with nonlinear and complexly intercoupled phenomena in a relatively short time and provide insight for quantitative understanding and better prediction. In the hands of enthusiastic and mature investigators, intractable problems will recede on a quickened time scale in this computationally synergized environment. (Zabusky, 1987)

Zabusky's prediction from 1987 has come true and this development of computational power has even resulted in new potential methods that could not have been predicted in the 1980s. A case in point is the success of using graphical processing units to run large scale simulations, a method which is much faster than multicore systems and vector processors.[21]

The reasoning style of computational science has permeated disciplines like meteorology, high energy physics, aerospace engineering and many other similar disciplines where non-linear processes and huge data sets are important. It is clear that this new style has already been very successful and has led to new promising scientific and engineering research programmes. But computational science is by nature an interdisciplinary activity and its success requires precise cooperation between several disciplines. However, the disciplines involved have very diverse perspectives and use very different methods. For instance, numerical analysis and computer science are two essential aspects of computational science which have contributed decisively to its development. All the same, computer science and numerical analysis have in a sense drifted apart. This was observed by Lenore Blum, Felipe Cucker, Michael Shub and Steve Smale in their important work *Complexity and real computation*:

> There is a substantial conflict between theoretical computer science and numerical analysis. These two subjects with common goals have grown apart. For example, computer scientists are uneasy with calculus, whereas numerical analysis thrives on it. On the other hand numerical analysts see no use for the Turing machine.
>
> ...
>
> A major obstacle to reconciling scientific computation and computer science is the present view of the machine, that is, the digital computer. As long as the computer is seen simply as a finite or discrete object, it will be difficult to systematize numerical analysis. We believe that the Turing machine as a foundation for real number algorithms can only obscure concepts. (Blum et al., 1998, p. 23)

[21]Cf. Tadmor (2012).

Our understanding of what scientific computing is or should be is not complete. There are several diverging developments. We need a better understanding of what a computation is over more complex structures than the natural number. In the case of natural numbers we have the Church-Turing Thesis which gives a precise definition of computation over the natural numbers, but we do not have a Church-Turing Thesis for computation in higher types or computation over more complex structures. In this area several new competing conceptions of computational structures have appeared, but they are very disparate and it is not clear how relevant they are for numerical analysis.

The work undertaken by Lenore Blum, Felipe Cucker, Michael Shub and Steve Smale is among the first attempts to develop a concept of computation which reflects the activities within fields where the new computational style is prevalent. It is not the final solution. However, we have every reason to expect that the close interaction between different fields within the new style of computational science will eventually lead to a new and better understanding of computation over complex structures.

References

Arnold, V. I. (2006). In V. I. Arnold, A. A. Bolibruch, Y. S. Osipov, L. D. Faddeev, Y. G. Sinai, V. B. Filippov, Y. I. Manin, V. M. Tikhomirov, & A. M. Vershik (Eds.), *Mathematical events of the twentieth century*. Berlin: Springer.

Aspray, W. (1990). *John von Neumann and the origins of modern computing*. Cambridge, MA: The MIT Press.

Atiyah, M. (2002). Mathematics in the 20th century. *Bulletin of the London Mathematical Society, 37*, 1–15.

Blum, L., Cucker, F., Shub, M., & Smale, S. (1998). *Complexity and real computation*. New York: Springer. With a foreword by Richard M. Karp.

Cantor, G. (1962). *Gesammelte Abhandlungen mathematischen und philosophischen Inhalts. Mit erläuternden Anmerkungen sowie mit Ergänzungen aus dem Briefwechsel Cantor-Dedekind.* Herausgegeben von Ernst Zermelo. Nebst einem Lebenslauf Cantors von Adolf Fraenkel. Hildesheim: Georg Olms Verlagsbuchhandlung.

Charney, J. G. (1955). Numerical methods in dynamical meteorology. *Proceedings of the National Academy of Sciences, 41*, 798–802.

Charney, J. G., Fjörtoft, R., & von Neumann, J. (1950). Numerical integration of the barotropic vorticity equation. *Tellus, 2*, 237–254.

Fermi, E., Pasta, J., & Ulam, S. (1965). Studies of non linear problems. In E. Segré (Ed.), *Colected papers of Enrico Fermi* (Vol. 2). Chicago: The University of Chicago Press.

Feynman, R. P. (1992). *The character of physical law*. London: Penguin Books.

Gödel, K. (1931). Über formal unentscheidbare Sätze der Principia Mathematica und verwandter Systeme, I. *Monatshefte für Mathematik und Physik, 38*, 173–198.

Hacking, I. (1992a). Statistical language, statistical truth and statistical reason: The self-authentication of a style of scientific reasoning. In E. McMullin (Ed.), *The social dimensions of science* (pp. 130–157). Notre Dame: University of Notre Dame Press.

Hacking, I. (1992b). 'Style' for historians and philosophers. *Studies in History and Philosophy of Modern Physics, 23*, 1–20.

Hammig, R. W. (1980). The unreasonable effectiveness of mathematics. *The American Mathematical Monthly, 87*(2), 81–90.

Kac, M., & Ulam, S. M. (1979). *Mathematics and logic*. New York: Penguin Books. Retrospect and prospects, Abridged reprint of the 1968 edition.
Lynch, P. (2008). The eniac forecast, a re-creation. *Bulletin of the American Meteorological Societ, 89*(1), 45–55.
Rees, E. G. (1988). *Notes on geometry*. Berlin: Spinger.
Schweber, S., & Wächter, M. (2000). Complex systems, modelling and simulation. *Studies in History and Philosophy of Modern Physics, 31*, 583–609.
Segré, E. (Ed.). (1965). *Collected papers of Enrico Fermi* (Vol. 2). Chicago: The University of Chicago Press.
Tadmor, E. (2012). A review of numerical methods for nonlinear partial differential equations. *Bulletin of the American Mathematical Society, 49*(4), 507–554.
Turing, A. M. (1937). On computable numbers, with an application to the entscheidungsproblem. *Proceedings of the London Mathematical Society, 42*, 230–265.
von Kármán, T. (1940). The engineer grapples with nonlinear problems. *Bulletin of the American Mathematical Society, 46*(8), 615–683.
Weisset, T. P. (1997). *The genesis of simulation in dynamics*. Berlin: Springer.
Wigner, E. P. (1967). The unreasonable effectiveness of mathematics in the natural sciences. In E. P. Wigner (Ed.), *Symmetries and reflections*. Bloomington and London: Indiana University Press.
Zabusky, N. J. (1987). Grappling with complexity. *Physics Today, 40*(10), 25–27.
Zabusky, N. J. (2005). Fermi-pasta-ulam, solitons and the fabric of nonlinear and computational science: History, synergetics, and visiometrics. *Chaos, 15*, 1–16.

A Fashionable Curiosity: Claudius Ptolemy's 'Desire for Knowledge' in Literary Context

Johannes Wietzke

Abstract

This article examines in detail the second-century CE polymath Claudius Ptolemy's expression of the 'desire for knowledge', situating it against a wider backdrop of similar expressions in the Greek textual tradition. I argue that in his expression, Ptolemy creatively alludes to Plato's *Phaedrus*, a practice that, surprisingly, here ties his work more closely to contemporary oratory and the 'novel' than to generic precursors in the exact sciences. The piece thus demonstrates how an author in the highly formalized genre of mathematics employs specific textual strategies held in common with his wider, contemporary literary culture.

I would like to thank the organizers of the Guangzhou conference for including me in the program, as well as for showing such wonderful hospitality over the course of the conference weekend. Acknowledgements are due, too, to audiences at Stanford and in Philadelphia who heard and commented on earlier versions of the paper. I am especially grateful to Geoffrey Lloyd, Reviel Netz, Anastasia-Erasmia Peponi, Susan Stephens, and the two anonymous referees, all of whom offered comments on earlier drafts that resulted in important clarifications and improvements. All errors of fact and interpretation remain my own.

J. Wietzke (✉)
Department of Classical Studies, Trinity University, Chapman Center, San Antonio, TX 78212, USA
e-mail: jwietzke@trinity.edu

S. Ju et al. (eds.), *Cultures of Mathematics and Logic*, Trends in the History of Science, DOI 10.1007/978-3-319-31502-7_5

> I frankly admit that I am strongly attracted by the simplicity and beauty of the mathematical schemes with which nature presents us. You must have felt this, too.
>
> —Werner Heisenberg to Albert Einstein[1]

The ancient Greek exact sciences constituted a literary practice. Therefore, we may interpret them as we interpret other expressions of Greek literary culture: through textual analysis sensitive to form and style, along with close readings in comparison with other literary texts inside and outside of generic limits. Such is my basic methodological assumption for this paper, and in this I follow a growing number of scholars who have been making an ever more compelling case that an appraisal of ancient mathematical and technical texts is essential for understanding the complexities of Greco-Roman literary culture as a whole.[2] In this paper, I specifically investigate the literary dimensions of how mathematical authors express their motivations for engaging in the practices of the exact sciences, focusing on the second-century CE polymath Claudius Ptolemy's expression of what we may call the 'desire for knowledge'.[3] My own desire in this paper is twofold: on the one hand, I wish to use this expression to locate Ptolemy in his literary context, both synchronically and diachronically, in genres both outside and inside of the exact sciences. On the other, I hope to show some of the ways in which the desire for knowledge, however essential it is to human nature, is expressed through context-dependent forms. Heisenberg's frank admission of attraction perhaps reflects a modern scientific culture that values matter-of-fact expression; in contrast, we shall see how Ptolemy's desire for knowledge is expressly charged with an eroticism that animated the literary ambitions of the second century CE.

1 Ptolemy Outside of Literary Context

Despite the perennial interest in revealing the scientific dimensions of Ptolemy's writings and their place and influence in the history of science, Ptolemy is still underserved in studies of ancient science that seek to situate those writings in their wider literary and social contexts. Partly this seems due to the very form of

---

[1]Heisenberg 1971: 69.

[2]An early example is Fuhrmann 1960, a formalist account of ancient handbooks, but the last decade and a half, especially, have witnessed an acceleration in shorter scholarly publications on technical and scientific writing. An important methodological essay is van der Eijk 1997, but there are still relatively few monographs that account for formal or otherwise literary aspects of ancient scientific texts, though Netz 1999 and 2009 and Asper 2007 are crucial contributions. See also Fögen 2009 for a linguistic approach to Roman technical writing, and Mattern 2008 on the form and rhetoric of medical narrative across Galen's works.

[3]The discussion here thus complements Matthew Leigh's 2013 study of curiosity that focuses on πολυπραγμοσύνη. I hope, too, that this philological investigation of mathematical desire will dovetail with recent philosophical discussions of one object of that desire, namely, mathematical beauty; cf. Rota 1997, and Müller-Hill and Spies 2011.

his mathematical enterprise: a text such as the *Mathematical Syntaxis* (hereafter, *Syntaxis*), replete with geometric proofs, tables, and diagrams, does not lend itself to easy comparison with works in other genres. Moreover, as Reviel Netz has argued, original and challenging work in the geometric sciences, exemplified by the proofs in Ptolemy's *Syntaxis*, was a marginal cultural practice even among elites in antiquity, and evidence suggests that this was especially true in Ptolemy's century.[4] Other practices with a more secure and obvious standing in ancient society have naturally garnered more attention from social historians and literary scholars. Medicine is the case in point, and the magnetic Galen has attracted the attention of scholars pursuing all manner of research question.[5] Partly, however, the general neglect of Ptolemy as a source on wider social and literary trends must be the product of his own silence about such trends. Widely recognized is the fact that, for one who wrote so much (amounting to some 300,000 words extant), Ptolemy writes surprisingly little that would clearly reveal details about his own life and cultural context. In the words of Alexander Jones: 'In [Ptolemy's writings] there is no personality, no reference to himself except as an observer, scholar, and theoretician, no allusion to his environment.'[6] Likewise we have no substantial attestations of his life or works from other sources prior to the commentaries of Pappus, Theon, and Proclus, which date to the fourth century and afterward.[7] Unlike so many of his contemporaries, Ptolemy appears never to have traveled, either to Rome or elsewhere, and we know nothing about his non-scientific activity in or around Alexandria, save for a possible connection to the temple of Serapis at Canopus.[8] Nor can we securely imagine his active involvement in the bustling social life of Alexandria itself, as a tantalizing though late (sixth-century) anecdote situates him in isolation for 40 years on the outskirts of the city.[9] In short, Ptolemy appears to have put minimal effort into creating a public image for himself; in this he is at the pole opposite to Galen.

The above merely gestures toward the serious challenges that interfere with our understanding of Ptolemy, both as a man and as a man of his time. In spite of Ptolemy's apparent detachment, however, recent studies have sought to ground him

[4]Netz 2002. On the apparently low mathematical productivity of the second century CE, see the tables in Netz 1997: 6–10. The *Syntaxis*, of course, is entangled to some degree with astrological practices of much wider popularity, but as Bernard 2010: 513 notes, even when an astrological theory was supported by geometric models, one did not need to comprehend those models in order to calculate horoscopes from, say, numerical tables.

[5]See the recent essays and up-to-date bibliography in Gill et al. 2009. To be sure, what we would call 'astrology' occupied a position arguably comparable to medicine, and Ptolemy's *Tetrabiblos* has been an important source for investigations of it as a social practice (e.g., Barton 1994: 27–94).

[6]Jones 2010: *xi*; cf. Toomer 1975.

[7]See Jones 1990, however, for papyrological evidence of early, perhaps even contemporary, criticism of Ptolemy's lunar theory. Toomer 1985: 204 argues that the sole mention of Ptolemy in Galen's works is an interpolation from the Arabic tradition.

[8]Cf. Jones 2005a: 62.

[9]Olymp. *In Phd.* 10.4, granted some plausibility by Jones 2005a: 61–64.

in the wider scientific, philosophical, and social contexts of the second century. A brief survey: Alexander Jones and Anne Tihon have identified papyrological evidence of theoretical astronomy, similar in form to (perhaps rivaling?) the *Syntaxis*, by Ptolemy's contemporaries and immediate predecessors.[10] Aiming to more precisely define Ptolemy's immediate reception, Alain Bernard has recently argued that the *Syntaxis* was written for an audience of astrological practitioners,[11] while Jacqueline Feke has situated Ptolemy in the midst of Middle-Platonic and Aristotelian debates on ethics and epistemology.[12] Cristian Tolsa has recently brought to the fore the social context in which these debates occurred.[13] Scholarship thus continues to develop an ever-higher resolution image of Ptolemy participating in contemporary intellectual practices. Still lacking, however, is a concerted effort to examine the literary aspects of Ptolemy's achievements and situate them in a literary context, according to the interpretive methods I listed at the outset.[14] I cannot achieve that in this short paper, of course, but my hope is that an analysis of one aspect of Ptolemy's literary craft, his expression of the 'desire for knowledge', will demonstrate the interest of the larger project.

2 Ptolemy's 'Desire for Knowledge'

Let us then turn to Ptolemy's account of that desire. To begin, we shall examine a passage from the third and final book of what is probably one of his earliest works, the *Harmonics*.[15] The passage marks a pivotal moment in the treatise, when Ptolemy both announces the fulfillment of the study's primary goal and reflects on its conclusion, which is that the principles of harmonics that he has determined by geometric methods conform to what can be determined through auditory perception. He claims to have demonstrated, in other words, that rationalist and empiricist methods operate in harmony with one another. Advancing toward the concluding part of the treatise, he then describes the twofold effect that reflection on the 'harmonic power' (ἁρμονικὴν δύναμιν)[16] induces, stating:

> Since it may follow for a person who has theorized on (θεωρήσαντι) these matters to be filled with wonder (τεθαυμακέναι) immediately—if he wonders also at other things of exceptional beauty—at the extreme rationality of the harmonic power, and at the fact that

[10] Jones 2004 and Tihon 2010.

[11] Bernard 2010.

[12] Feke 2009 and 2012.

[13] Tolsa 2013.

[14] Moving in this direction are Jones 2005b, which patiently maps out several of Ptolemy's labyrinthine rhetorical strategies, and Mansfeld 1998: 66–75 and Feke 2012: 89, which consider some formal aspects of Ptolemy's prologues. Tolsa 2013: 301–328 situates 'Ptolemy's epigram' (possibly dubious but transmitted in the manuscript tradition of the *Syntaxis*) in literary context.

[15] For a plausible sequence of Ptolemy's authorship, see Feke and Jones 2010: 200–201.

[16] For different interpretations of what, exactly, the harmonic power is, see Barker 2000: 259–263, Swerdlow 2004: 151 and Feke 2009: 69–91.

> it finds and creates with perfect precision the differences of its own forms; and, on the other hand [since it may follow for him], owing to some divine love (ὑπό τινος ἔρωτος θείου), to desire (ποθεῖν) to behold, as it were, the nature of [the harmonic power] (τὸ γένος αὐτῆς ὥσπερ θεάσασθαι) and with what other things it is conjoined among the things comprehended in this world, we shall try, in summary fashion and so far as we are able, to investigate this remaining part of our theoretical undertaking, to display the magnitude of this kind of power.[17]

Untangling Ptolemy's syntax, we see that the very act of theorizing provokes two emotional reactions in the individual: wonder and desire. Ptolemy's feeling of wonder as it pertains to the activity of theory deserves its own discussion.[18] Our present focus, however, centers on his desire, and let me stress the evocative language with which he expresses it: the theorist of harmonics (whom we may understand both in a general sense and as Ptolemy and his reader) is in the grip of a divine love. That word for 'love' (ἔρωτος) especially grabs our attention, as it specifies that this is erotic passion. Moreover, the theorist desires (ποθεῖν) to behold (θεάσασθαι) both the nature (γένος) of the harmonic power and what it is conjoined with in this world. In other words, he desires to know both what the harmonic power is and how to classify it. His desire is thus the desire for knowledge. Moreover, we find similar sentiments in the *Syntaxis*. In the preface to that work, Ptolemy further describes the erotic attraction of the exact sciences, this time astronomy: therein he claims his intellectual mission as to increase 'the love (ἔρωτα) of contemplating (θεωρίας) the eternal and unchanging',[19] which is to say, the love of contemplating celestial bodies through mathematics. Moreover, Ptolemy describes those, like himself, who pursue mathematical astronomy as 'lovers (ἐραστάς) of divine beauty',[20] characterizing them, again, not by simple affection but by erotic passion. All of these evocative expressions frame Ptolemy's attitude toward investigation in the exact sciences as a desire for knowledge.

Moreover, the language that Ptolemy uses to describe that desire is not only evocative, but also (for Ptolemy, at least) rare: nowhere else in the *Harmonics*, for instance, do we find inflections of ἔρως, ποθέω or θεάομαι. Indeed, the passage cited above features Ptolemy's only use of ποθέω in the whole of his corpus. Ptolemy uses θεάομαι only once more, in the *Syntaxis*, but in what appears to be an otherwise lexically uninteresting discussion of epicycles.[21] On the other hand, ἔρως and its cognates occur a total of 13 times throughout Ptolemy's writings. Most of these, however, occur in the course of the technical discussion in the *Tetrabiblos* and concern the general affections of character variously wrought by the combinations of celestial bodies; they pertain to the motivation of neither the author nor a general figure of the astrologer. Thus we arrive at another important point: in addition to

[17] Ptol. *Harm.* III.3 [92.1–8 Düring], translation adapted from Barker 1989: 371.

[18] For a general account of this issue in Greek thought and literature up to (pseudo-)Longinus (not including Ptolemy), see Nightingale 2004: 253–268.

[19] Ptol. *Syntaxis* part 1, p. 7.25–26 Heiberg.

[20] Ptol. *Syntaxis* part 1, p. 7.21–22 Heiberg.

[21] Ptol. *Syntaxis* part 1, p. 361.11 Heiberg.

being exceptional in their lexical form, expressions of the desire for knowledge are also confined to certain parts of Ptolemy's work. We find them only in extended 'second-order' passages, in which Ptolemy reflects self-consciously on methods, practices and, as we have seen, motivations, and which function as an introduction to or transition between extended expositions on 'first-order' (i.e., technical or theoretical) material.

3 The Desire for Knowledge in Greek Literature: A Survey

Such expressions of the desire for knowledge are thus rare in Ptolemy's works, but is the same true in Greek textual culture leading up to the second century CE? To gain an impression of how unique Ptolemy is in this regard, I conducted a series of correlated searches using the online Thesaurus Linguae Graecae (TLG). Here let me offer a few words in the way of methodology: I began with the key terms that Ptolemy himself uses that correspond to 'desire' and 'knowledge', but added as well cognate nouns and verbs (and one adjective), as well as additional nouns and verbs that seemed reasonable to include so as to further round out the impression; the full list is presented below (words in bold are found in Ptolemy's own combinations)[22]:

'Desire': **ἔρως/ἐραστής**/ἐράω/ἔραμαι/ἐρωτικός, πόθος/**ποθέω**, ἐπιθυμία/ἐπιθυμέω, ὄρεξις/ὀρέγω, ἵμερος/ἱμείρω, βούλησις/βούλομαι

'Knowledge': **θεάομαι**, **θεωρία**/θεωρέω, οἶδα, ἀλήθεια, γνῶσις/γιγνώσκω, ἐπιστήμη, μανθάνω, σοφία, πυνθάνομαι

Although I have grouped the terms under two general headings, I make no claims about synonymity within each group. Quite on the contrary, both groups obviously include terms that signify a range of concepts and activities, and even individual words may feature different nuances from one author or text to the next. But this should not matter for our purposes: a combination of any item from the first set with any item from the second generally conveys or pertains to the 'desire/desirer/desiring for knowledge/knowing'. Moreover, in this literary study, again, we are more concerned with the various expressions that signify a general concept, rather than defining precisely what the signified concept is. Other terms, too, certainly could have been included to present a more exhaustive picture of that

[22]To set out my search methodology more precisely: I used the TLG's 'Advanced Lemma Search' function to locate combinations of each of the given terms for desire and knowledge occurring within one line of each other. For the 'desire' group, no formal constraints were set (on case, number, tense, mood, etc.), so as not to bias the search results to favor certain parts of syntax. On the other hand, since terms of knowledge are here specifically the objects of desire, certain constraints were set for this group in all searches: all noun-searches were limited to the singular genitive and accusative; verb-searches were limited to present and aorist, active infinitives, except for the cases of οἶδα, where I targeted the perfect, active infinitive, and the deponent πυνθάνομαι, where I targeted the present and aorist, middle/passive infinitive.

variety of expression. But this is foremost a study of Ptolemy, not the 'desire for knowledge' itself.[23] Thus limiting the scope to Ptolemy's own combinations and the combinations built from the other terms in each set should allow us to reach reasonable conclusions about Ptolemy's uniqueness of expression, while providing a good background-impression of other possibilities.

Just as my search aims to produce a representative, rather than exhaustive, pool of evidence, so does my ensuing discussion of that material consist of an impressionistic account of distributions especially sensitive to genre and chronology. My findings so far suggest that a more detailed analysis of the 'desire for knowledge' may tell us a lot about how culture shapes this apparently natural feature of human psychology, but to attempt this here would distract from our intended focus on Ptolemy. The present discussion instead adheres to three principles of discursive economy: first, only single instances of a particular combination (without regard for grammatical inflection) are noted for a given author, rather than a list of total counts per author or work. Tallying frequencies of each combination seems less significant for developing an impression of the temporal and generic 'reach' of an expression; the more qualitative approach I follow should bring absences and presences into conspicuous relief along those lines. Second, to frame the discussion with a simple and intuitive structure, I group the individual species of desire[24] into three genera, whose organizational logic pertains to the species' temporal and generic presence. In short, these are the common, the philosophical, and the Platonizing, and we shall examine each in turn. Third, I will be selective in my description of each genus, offering explanation only where it will be helpful for the ensuing discussion of Ptolemy.

There is no question that Greek authors leading up to Ptolemy conceived of a desire for knowledge; in what forms then did they express it? We proceed with the 'common genus'. First, a clarifying point: in my usage, 'common' indicates a wide temporal and generic distribution, not a high frequency of occurrence—though it was the case that some of these searches produced the highest returns.[25] This genus is the largest of the three, in fact the only one which encompasses more than one species of the desire for knowledge. These are πόθος/ποθέω, ἐπιθυμία/ἐπιθυμέω, ἵμερος/ἱμείρω, and βούλομαι, and they are represented in Athenian tragedy and comedy; fourth-century oratory and philosophy; the Classical and Hellenistic historians and geographers, and various prose works of the Imperial period, including our passage from Ptolemy's *Harmonics*.[26] The genus is thus indeed common, but also

[23]Leigh 2013 develops an intellectual history of ancient curiosity which is naturally sensitive to form, but whose primary focus is on the valence of πολυπραγμοσύνη and related concepts.

[24]I treat cognate nouns, verbs, and adjectives (e.g., πόθος and ποθέω) as a single species, locating the defining features thereof at the linguistic root.

[25]Expressions based on βούλομαι number in the dozens, whereas I count only seven examples that feature ἵμερος/ἱμείρω. Uniquely in my searches, βούλησις produced zero results.

[26]The following is a representative, not exhaustive list of citations: ΠΟΘΟΣ/ΠΟΘΕΩ: θεάομαι: Ph. *Jos*. 204; Plu. *Demetr*. 6.5; Ptol. *Harm*. 3.3.ἀλήθεια: Ph. *Aet*. 2; S.E. *M*. Pr.6. οἶδα: S. *Tr*. 632; E. *IT*. 542; Pl. *Men*. 84c; Arist. *PA*. 644b26; Str. 2.5.18; Luc. *Icar*. 4. θεωρία: Th. 6.24.3.

textured: as we might expect, certain terms in the 'knowledge' set (i.e., ἀλήθεια, ἐπιστήμη) appear exclusively in philosophical or quasi-philosophical discourses. It is also not all-pervasive: we notice particular absences in poetry other than Athenian drama, and it does not feature prominently in pre-Imperial medical discourse or the fragments of Hellenistic philosophy.[27] Most glaring for us is the absence of the expression from the exact sciences. All of these absences may simply be an effect of the survey's limited scope, however, and later in this essay I will examine more closely texts in the exact sciences as we try to understand Ptolemy's generic position.

The second genus, the 'philosophical', solely entails the desire-complex ὄρεξις/ὀρέγω. It appears less frequently overall, and then only in texts of a philosophical or quasi-philosophical nature (e.g., Nicomachus' *Introductio arithmetica*, Theon of Smyrna's *De utilitate mathematicae*, and the Galenic corpus; I count only one exception in Dionysius of Halicarnassus' *Roman Antiquities*).[28] An array of

---

θεωρέω : not found. ἐπιστήμη: Ph. *Op*. 77; Plu *Adv.Col.* 1118b. γνῶσις: Str. 13.1.1; Plu. *Adv.Col.* 1118b. γιγνώσκω: Ph. *Virt*. 215; Plu. *De genio* 590a; Gal. *MM*. K vol. 10, p. 714.17. μανθάνω: E. *Ion* 1432; Ph. *Fug*. 8; D.H. 7.66.1; Max.Tyr. 8.4; S.E. *M*. 9.75; Luc. *Trag*. 209. σοφία: Ph. *Op*. 5; Max.Tyr. 25.1; Luc. *Merc.cond.* 25. πυνθάνομαι: Plu. *Quaes.Rom.* 266b. ΕΠΙΘΥΜΙΑ/ΕΠΙΘΥΜΕΩ: θεάομαι: Pl. *Ti*. 19b; Arist. *Rh*. 1370a26; D.S. 13.9.3; Ael. *NA*. 16.39; Gal. *PHP*. 5.7.48. ἀλήθεια: Gal. *MM*. K vol. 10, p. 457.14. οἶδα: Ar. *V.* 86; Pl. *Grg*. 474c; Luc. *VH*. 2.20. θεωρία: not found. θεωρέω: Pl. *Lg*. 951a; Epicur. *Ep. ad Pythoclem* 94. ἐπιστήμη: Arist. *Pol.* 1288b17; Gal. *Quod animi mores* K vol. 4, p. 772.3. γνῶσις: D.H. 11.36.1. γιγνώσκω: LXX *Is*. 58.2; D.S. 10.8.3; Gal. *Loc.Aff.* K vol. 8, p. 144.7. μανθάνω: Ar. *Nu*. 656; Pl. *Hp.Mi*. 369d; X. *Cyr*. 4.3.15; Arist. *Rh*. 1371a32; Erot. 29.8; Plu. *De Pyth*. 395e; App. *Pun*. 430. σοφία: Ar. *Nu*. 412; Pl. *Phd*. 96a; LXX *Wi*. 6.20; S. 15.1.64; J. *Ap*. 1.111. πυνθάνομαι: Ar. *Lys*. 486; Is. 3.8; Plu. *De comm.* 1066d.

ΙΜΕΡΟΣ/ΙΜΕΙΡΩ: θεάομαι: Ph. *Praem*. 39; Ael. *NA*. 11.17.ἀλήθεια: S.E. *M*. 1.42. οἶδα: not found. θεωρία: not found. θεωρέω: not found. ἐπιστήμη: not found. γνῶσις : not found. γιγνώσκω: not found. μανθάνω: S. *Fr.* 314.134 Radt; Plb. 14.Pr.4; Ph. *Cont*. 75. σοφία: Ph. *Spec.leg*. 1.50. πυνθάνομαι: not found.

ΒΟΥΛΟΜΑΙ: θεάομαι: Ar. *Th*. 234; Pl. *R*. 327a; Plb. 7.12.1; D.S. 17.116.5; Ael. *VH*. 14.17; Plu. *Cat.Ma*. 17.4; Gal. *AA* K vol. 2, p. 630.8. ἀλήθεια: not found. οἶδα: Hdt. 1.86; Hp. *Aph*. 5.59; E. *Alc*. 140; Ar. *Nu*. 250; Th. 1.52.2; Pl. *Lg*. 629c; X. *Cyr.* 8.4.11; Isoc. 17.9; D. 19.227; Aesch. 3.199; Arist. *EE*. 1216b22; Plb. 4.38.12; D.H. *Dem*. 50; J. *AJ*. 1.325; D.Chr. 4.67; Gal. *Dig.puls*. K vol. 8, p. 955.17. θεωρία: not found. θεωρέω: Th. 5.18.2; Alc. *Od*. 86; D. *Ep*. 4.5; Arist. *Cael*. 300b20; D.S. 19.52.4; Gal. *Dig.puls*. K vol. 8, p. 944.8. ἐπιστήμη: not found. γνῶσις: not found. γιγνώσκω: S. *Fr.* 1130.3 Radt; Pl. *R*. 572b; X. *Mem*. 1.2.42; Arist. *De an*. 402a14; LXX *To*. 5.14; Plb. 1.1.5; D.S. 5.77.3; J. *AJ*. 12.100; D.Chr. 31.38; Gal. *Lib.prop*. K vol. 19, p. 9.1; Aesop. 50.6. μανθάνω: S. *Ph*. 233; E. *El*. 229; Ar. *Nu*. 239; Pl. *Sph*. 232d; X. *HG*. 6.5.52; D. 23.2; Agatharch. 14.6; Plb. 21.41.5; Str. 2.5.43; Ph. *Jos*. 56; D.H. 4.66.1; J. *BJ*. 7.454; Plu. *Crass*. 28.4; Ruf. *Syn.puls*. 3.3; Vett.Val. 8.8; S.E. *M*. 8.87; Ael. *NA*. 5.42; Gal. *PHP* 1.6.4; Luc. *VH*. 1.5. σοφία: not found. πυνθάνομαι: Hdt. 6.69; E. *Hipp*. 910; Ar. *Nu*. 482; Th. 8.19.1; Pl. *La*. 191d; X. *Oec*. 7.2; D. 4.10; Arist. *Top*. 161b5; Plb. 11.28.11; Plu. *De genio* 577e; S.E. *P.* 2.211; Gal. *De semine* K vol. 4, p. 527.18.

[27]I count only two instances of βούλομαι εἰδέναι in the Hippocratic corpus: *Aphor.* 5.59 and *De semine* 13.23, and one instance of ἐπιθυμέω θεωρεῖν at Epicur. *Ep. ad Pythoclem* 94.

[28]ΟΡΕΞΙΣ/ΟΡΕΓΩ: θεάομαι : not found. ἀλήθεια: Pl. *R*. 485d; Plu. *De recta* 48c; Ptol. *Judic*. p. 5 Lammert; Gal. *De const. artis* K vol. 1, p. 244.16. οἶδα: Arist. *Metaph*. 980a21. θεωρία: Alcin. *Intr.*

Classical and Imperial authors are represented within the relatively small set of results, but the expression does not seem to have been in high favor. I have found only one instance in Plato and none in Philo of Alexandria, both of whom are well-represented in the other genera, and any 'desire for knowledge', at least expressed in the terms of this study or the fragments we have, does not feature in the discourse of the Hellenistic philosophical schools.

Our last genus, the 'Platonizing', is composed only of those expressions of desire defined by ἔρως and its cognates.[29] A seemingly more descriptive label for this genus might be 'Imperial', as in fact most instances occur in a range of prose works beginning with those of Philo of Alexandria; in addition to Philo's hybrid writings, we find a high amount of oratory, medical writing, a Platonic handbook, one Plutarchian life, and only now, with Ptolemy, texts in the exact sciences. Prior to Philo we find only four instances of the genus: two from the Platonic corpus and one each from Sophocles and Euripides.[30] What then justifies the Platonizing label? The tragedians deny Plato any claim to inventing the eroticization of knowledge,[31] and Plato was in any case not the first to conceptualize ἔρως in terms abstract from personal relationships.[32] But as is plain from such dialogues as the *Phaedrus* and *Symposium*, developing the philosopher and his pursuit of knowledge in expressly erotic terms was a defining aspect of Plato's philosophical project.[33] Moreover, a TLG survey of Plato's use of ἔρως and its cognates reveals other knowledge-terms subject to them,[34] and further TLG searches confirm that an erotic desire for these is exclusive to Plato's dialogues until Philo.[35] Hence there does seem something definitively Platonic about the erotic desire for knowledge. And for now let me propose that its implementation by Imperial authors was a consequence of

27.4. θεωρέω: not found. ἐπιστήμη: Arist. *De an.* 433a6; Gal. *Syn.puls.* K vol. 9, p. 431.2. γνῶσις: D.H. 1.1.3; Theon Sm. 1.11. γιγνώσκω: Gal. *Diff.resp.* K vol. 7, p. 889.1. μανθάνω: Gal. *Ars med.* K vol. 1, p. 224.3; Cels. *Apud Originem* 6.18. σοφία: Nicom. *Ar.* 1.2.3; Alcin. *Intr.* 1.1; Gal. *MM.* K vol. 10, p. 114.18. πυνθάνομαι: not found.

[29]ΕΡΩΣ/ΕΡΑΣΤΗΣ/ΕΡΑΩ/ΕΡΑΜΑΙ/ΕΡΩΤΙΚΟΣ: θεάομαι: Ph. *Praem.* 38; Ptol. *Harm.* 3.3 (with ποθεῖν). ἀλήθεια: Pl. *R.* 501d; Ph. *Spec.leg.* 1.59; Alcin. *Intr.* 1.2; Max.Tyr. 16.2; Gal. *Nat.fac.* K vol. 2, p. 179.13; Ael. *NA.* 2.11. οἶδα: not found. θεωρία: Ptol. *Synt.* 1.7; Gal. *Dig.puls.* K vol. 8, p. 860.5. θεωρέω: not found. ἐπιστήμη: Pl. *Ti.* 46d; Ph. *Op.* 77; Thess. *Virt.herb.* Pr.5. γνῶσις: not found. γιγνώσκω: not found. μανθάνω: E. *Hipp.* 173; Max.Tyr. 11.11. σοφία: Ph. *Op.* 5; Plu. *Sol.* 2.2; D.Chr. 36.40; Max.Tyr. 18.5; Ael. *NA.* Ep.1. πυνθάνομαι: S. *OC.* 511.

[30]Another might be found at E. *Fr.* 889.1 Nauck: παίδευμα δ'Ἔρως σοφίας, but in context σοφίας is better construed quasi-subjectively with παίδευμα, rather than as the object of Ἔρως.

[31]Could the precedents from tragedy be yet further evidence of Plato's appropriation of poetic discourse for his construction of philosophy, as argued by Nightingale 1996?

[32]ἔρως is widely conceived as a passion for *polis* and power in fifth-century poetry and prose. For discussion, see Cornford 1907: 201–220; Arrowsmith 1973; Rothwell 1990: 37–43; Connor 1992: 96–98; Nightingale 1996: 187–188.

[33]The implications of this are explored in Halperin 1985. Cf. Nightingale 1996: 128–129.

[34]μάθημα (*R.* 485b), τὸ ὂν (*R.* 501d), νοῦς (*Ti.* 46d), φρόνησις (*Phd.* 68a), τὸ ἀληθές (*Phlb.* 58d).

[35]The question of why other authors apparently avoid this complex of expressions until the Imperial period cannot be answered here; more searches are warranted and may of course qualify the result.

the centrality of Plato's texts to Greek literary culture and education, as well as to Plato's otherwise well-documented influence on individual works by authors such as Philo, Maximus of Tyre, and Galen (which is not to reduce any of these to being mere 'Platonists').[36] Thus Imperial authors who render the desire for knowledge in erotic terms draw it closer to its presentation in a celebrated, Classical source; they Platonize it.

4 Ptolemy's Choice

The foregoing survey displays some of the range and texture that characterize the expression of the desire for knowledge in the Greek literary tradition through the second century CE. To refocus the discussion on Ptolemy, the key point is that to him as an author, that tradition offered a variety from which he could *choose*.[37] By choosing to cast that desire in consistently Platonizing terms, Ptolemy acts entirely in keeping with the Platonic philosophical currents that influence his texts,[38] as well as with the literary practice of other writers of his era. Noting the apparent expectedness of this choice, we might question how interesting it can even be—is Ptolemy's desire for knowledge, even in its erotic fervor, nothing more than a literary *topos*? Here we turn to literary criticism for guidance. In his study of allusion and intertextuality in Augustan poetry, Stephen Hinds observes that '*topos*' hardly constitutes an 'inert category': a Virgil does not simply insert a poetic commonplace pre-fabricated into his verses, but creatively reconfigures it, varying details and the manner of expression, thus transforming it into something 'new and fresh'.[39] Could Ptolemy be doing something similar?

Ptolemy's expressions of desire in the *Syntaxis* are perhaps too brief to yield much interpretive fruit; instead we shall focus on his extended statement in the *Harmonics*. Let me offer again the key phrase: 'It may follow for him who has theorized on these matters, owing to some divine love (ὑπό τινος ἔρωτος θείου), to desire (ποθεῖν) to behold, as it were, the nature of the harmonic power (τὸ γένος αὐτῆς ὥσπερ θεάσασθαι).' Even if the desire for knowledge expressed

[36]See De Lacy 1974 and now Hunter 2012 for Plato's influence on Imperial literary culture, and Trapp 1990 on the particular prominence of Plato's *Phaedrus*, one of the key sources for Platonic ἔρως. Mansfeld 1994: 58–107 describes the pedagogical context. For Platonic influence on Philo, Maximus, and Galen, see Dillon 1977: 139–183; Trapp 1997: *xxii–xxxii*; and De Lacy 1972.

[37]The fact that he made *any* positive choice reveals something about the generic history of the exact sciences: we noted the apparent absence of the desire for knowledge from any Hellenistic mathematical works. As our survey was admittedly limited, we shall examine these texts in more detail below.

[38]See, for example, the editorial notes on *Harm.* 3.3–5 in Barker 1989: 373–377; Taub 1993, esp. 31–34; Feke 2009; Feke 2012; Tolsa 2013, esp. chapters 1–3. Like other Imperial writers, of course, Ptolemy is not swept entirely away by those currents, but demonstrates a notable degree of eclecticism; Feke 2009: 221 brands his philosophy 'Platonic empiricism'.

[39]Hinds 1998: 40.

here is merely a *topos*, that consideration requires two qualifications in view of Hinds. First, Ptolemy is not quoting anyone. The exact form of the statement is not formulaic nor found elsewhere, but is Ptolemy's own. Second, Ptolemy reconfigures the key terms ἔρως, ποθέω, and θεάομαι in a unique manner: a TLG search produces Philo of Alexandria's description of an 'immense desire (ἔρωτι) to see (θεάσασθαι)', immediately qualified as a πόθον, as the only precedent for those terms in such close composition.[40] Unlike Philo, who in effect equates the two desires, Ptolemy distinguishes ἔρως from the action of πόθος by rendering the former as the source of the latter. At least in its expression, then, Ptolemy's desire for knowledge is conceptually more developed than that of other authors. If the 'desire for knowledge' is a *topos*, Ptolemy's is not typical.

5 Ptolemy's Platonic Enthusiasm

Indeed, by developing the expression so vividly in the *Harmonics*, Ptolemy can be seen to be making a specific, textual allusion to Plato's *Phaedrus*. That Ptolemy should refer to Plato at all is not surprising, given the Athenian philosopher's contemporary prominence and, more specifically, his clear influence on Ptolemy.[41] Up to now, however, this connection has received little attention from scholars who have pursued other questions.[42] But the passage is significant, for as will be made clear, the allusion to the *Phaedrus* situates Ptolemy in an unexpected literary context, and the nature of the allusion itself suggests something about his authorial aspirations. Let us then examine how this allusion works.

The allusion operates on two levels: first, there is a lexical correspondence between key terms from the *Harmonics* and the *Phaedrus* (in the absence of any explicit naming of Plato or the dialogue, this is how we recognize the allusion at all). The verb ποθέω is used four times in the dialogue: once by Socrates to invite Phaedrus to ask if he desires anything (*Phdr.* 234c); twice to describe the mutual desire of the lover and the beloved to be near each other (*Phdr.* 255d); and once to describe a soul 'full of desire' (ποθοῦσα) that races wherever it hopes to 'see' (ὄψεσθαι) a beautiful boy (*Phdr.* 251e). Forms of θεάομαι occur five times in the dialogue, denoting the beholding of various objects: true things (*Phdr.* 247e); the earthly namesake of beauty (*Phdr.* 250e); men of certain classes (*Phdr.* 271d); the writings of great writers (*Phdr.* 258c); and, most important for our discussion, the nature (γένος) of the earthy imitation of the forms of justice, temperance, etc. (*Phdr.*

[40]Ph. *Praem.* 38. The result followed TLG Advanced Lemma Searches for combinations of ἔρως/ἐραστής/ἐράω/ἔραμαι/ἐρωτικός, πόθος/ποθέω, θεάομαι occurring within one line of each other. After Ptolemy's *Harmonics*, the next instance is Ps.-Luc. *Am.* 53, perhaps from the early fourth century.

[41]See nn. 36 and 38 above.

[42]Tolsa 2013: 84 briefly notes lexical correspondence between the *Harmonics* passage and the *Phaedrus*.

250b). The various inflections of ἔρως are used too frequently to account for all instances here, but we may note that Socrates does call it 'a god or something divine' (*Phdr.* 242e: θεὸς ἤ τι θεῖον ὁ Ἔρως).

Here I offer two observations: first, beyond the basic correspondence of individual words between the two texts, we also find several instances of correspondence between word-groupings: thus, from the passages surveyed above, compare Socrates' statement in the dialogue that if not a god, '*erôs* is something divine' (τι θεῖον ὁ Ἔρως) with Ptolemy's description of 'divine *erôs*' (ἔρωτος θείου) in the *Harmonics*. Moreover, there is a strong correspondence between the statement in the *Phaedrus* that souls 'behold the nature' (θεῶνται τὸ ... γένος) of the earthly imitations of the forms of justice, etc., and that found in the *Harmonics* concerning the theorist's desire to 'behold the nature' (τὸ γένος ... θεάσασθαι) of the harmonic power (nor should we overlook the fact that in both the *Phaedrus* and the *Harmonics*, the objects whose nature is being regarded—on the one hand, the forms of justice, temperance and the like, and on the other hand, the harmonic power—are objects of pronounced beauty[43]). Lastly, while there is not the same, precise lexical correspondence in the verbs of sense-perception, the statement in the *Phaedrus* describing the soul 'full of desire' (ποθοῦσα), racing wherever it hopes to 'see' (ὄψεσθαι) a beautiful boy, overlaps semantically with the phrase found in the *Harmonics* concerning the theorist "desiring to behold" (ποθεῖν ... θεάσασθαι) the nature of the harmonic power.

The second observation is less obvious but critical for the present analysis: except for three exceptions,[44] all of the terms in the *Phaedrus* that correspond to those in the *Harmonics* occur in Socrates' second speech to Phaedrus (*Phdr.* 244a–257b).[45] There he vividly describes the philosophical lover whom others think simply mad, but whose madness is in fact inspired by a divine love (ἔρως). Socrates tells us that prior to the lover's present life, his soul had caught a glimpse of true beauty in the course of its heavenly flight. Now embodied in the lover, it beholds the nature of beauty and the concepts of justice and temperance, all manifest in young boys (*Phdr.* 250b). Recollecting these true forms, the soul of the lover, now full of desire, races toward wherever it hopes to see them (*Phdr.* 251e).

The account, brief as it is, should strike a familiar chord after the preceding comparison between the *Phaedrus* and *Harmonics*. The same passages that comprise the account of Plato's madman-lover are the same that were analyzed in the discussion of lexical overlap, and my conclusion is by now obvious: I submit that Ptolemy has shaped his theorist of harmonics—under the influence of divine love, desiring to behold the nature of the harmonic power—specifically in the image of Plato's inspired lover. This, then, is the second level of Ptolemy's allusion: the passage in

[43]See Feke 2009: 91–97 and Barker 2010 on beauty in Ptolemy's *Harmonics*.

[44]Socrates asking whether Phaedrus 'desires' anything (*Phdr.* 234c); 'beholding' the writings of great writers (258c) and men of certain classes (271d).

[45]Or, in the case of 'divine *erôs*' alone, in Socrates' recantation, which introduces the speech.

the *Harmonics* evokes not only Platonic language, but also a uniquely conceived character from a particular dialogue.

6 The *Phaedrus* in Greek Imperial Literature

The choice of dialogue is itself significant in view of Ptolemy's historical context. In the second century the *Phaedrus* exerted an especially powerful influence over textual culture.[46] The text itself was widely read, providing a model for literary style, and the dialogue inspired explicit engagement on the part of orators, writers of narrative fiction, and, no less, Galen.[47] Thus the impression of Michael Trapp:

> It must have been hard for the *pepaideumenos* to emerge from his education, whether rhetorical, philosophical, or both, without having been invited to study and admire this dialogue, and without having come to regard it as a proper model for imitation in his own literary products.[48]

The soul and passions of the madman-lover was itself a popular motif for imitation and allusion.[49] What is therefore interesting about its presence in the *Harmonics* is that it nudges Ptolemy out of the apparent social isolation of the exact sciences and into company with sophists, fiction-writers, and learned doctors.[50] Ptolemy's choice to allude to the dialogue's madman-lover, then, may reveal more about him as a prose-stylist than as an authority on harmonic theory.

To deepen our understanding of how the literary potential of the *Phaedrus* was realized in the second century, let us consider how a contemporary prose-stylist uses its language of eroticized psychology. Here we turn to Longus, author of the narrative fiction we know as *Daphnis and Chloe*.[51] The work opens with a narrative conceit, itself reminiscent of the *Phaedrus* in its detail and narrative timing: in a prologue, the narrator describes how inspiration to write the story seized him while

[46]A culture commonly referred to as the 'Second Sophistic', on which see Whitmarsh 2005.

[47]See De Lacy 1974: 6–8 and especially Trapp 1990, which includes an appendix that registers allusions to the *Phaedrus* (including the 'soul of the lover') from numerous authors and works; Ptolemy is not included. Cf. Hunter 1997; Hunter 2012: 151–184; and Rocca 2006 on Galen. Papyrological records from the CEDOPAL database (http://www2.ulg.ac.be/facphl/services/cedopal/), accessed on February 17, 2013, corroborate the dialogue's popularity in the Imperial period: out of 105 fragments of Plato, 8 are from the *Phaedrus* (three of which, moreover, feature text from Socrates' second speech). This number is exceeded only by the *Republic* (13), *Phaedo* (11) and *Laws* (10), but the *Phaedrus*' tally is especially impressive given the substantially greater length (and therefore, the greater odds of survival) of those other works.

[48]Trapp 1990: 141.

[49]For a sizeable list of allusions, see Trapp 1990: 172; to this we might add Gal. *Nat.fac.* K vol. 2, p. 179.12–15.

[50]Feke and Jones 2010: 200–201 situate the *Harmonics* early in Ptolemy's career; might its author be Trapp's young *pepaideumenos*, eager to demonstrate his rhetorical-philosophical education? Tolsa 2013: 201–203 contends that Ptolemy frames another early work, *On the Criterion*, as a rhetorical exercise.

[51]On date, author, and title, see Hunter 1983: 1–15.

hunting in the picturesque, natural setting of a *locus amoenus*.[52] There he happened upon a richly evocative painting, the details of which he describes to the reader in a tasting menu of sorts, featuring samples of the narrative delights that follow in the course of the work:

> Women giving birth, others dressing the babies, babies exposed, animals suckling them, shepherds adopting them, young people pledging love, a pirates' raid, an enemy attack—I gazed in wonder (ἰδόντα με καὶ θαυμάσαντα) at many other things, all of them erotic (ἐρωτικά), and a desire (πόθος) took hold of me to write in response to the painting (ἀντιγράψαι τῇ γραφῇ).[53]

Passing over the alluring details of the painting, we focus instead on the narrator's professed reaction to them. Like Ptolemy, Longus' narrator describes his motivation as desire operating in tandem with vision and erotic love (and wonder, no less). Their conceptual vocabulary of motivation is thus essentially the same, essentially Platonic. That Longus presents those concepts in a different permutation does not undermine this.[54] The point, again following Hinds, is that allusion is not numb imitation, but entails some degree of creative refashioning. Moreover, in Longus and other second-century authors, this refashioning entails not only introducing evocatively Platonic language and concepts into new literary contexts, but sometimes actually redefining what those concepts are, even to the point of contrasting them directly with the Platonic original. In the above passage, Longus appears to engage with Plato in a critical manner: Richard Hunter has argued that the narrator's 'writing in response to the painting' in fact gives voice to the painting's silence, thus the act tacitly contends with Socrates' claim—again, from the *Phaedrus*—that all writing and painting must remain forever mute.[55] Similar reconfigurations of Plato can be found in other authors of the period such as Plutarch and Maximus of Tyre.[56]

7 Ptolemy's Literary Ambition

Ptolemy's choice of the *Phaedrus* is accompanied by a similar degree of literary ambition. The allusion is more than an inert, merely imitative reference to Plato's madman-lover. It effectively reconfigures not only the expression of the *topos* of the desire for knowledge, but also the very conception of the Platonic lover. Note, first, that Ptolemy signals a figurative turn in the discourse with syntactical and lexical cues. Although ostensibly making Platonic desire the logical result of theoretical activity, Ptolemy renders the main verb of the clause in the potential-optative: 'Since it *may* follow (ἀκόλουθον ἂν εἴη) for one who has theorized on these matters . . . '

[52]On the Platonic aspects of this opening, see Hunter 1997: 24.

[53]Longus Pr. 2, translation adapted from Reardon 1989: 289.

[54]In Longus' account, for instance, vision prompts desire, whereas for Ptolemy the desire is to see.

[55]Hunter 1997: 28.

[56]See, respectively, Whitmarsh 2001: 47–57 and Tarrant 2000: 133–135.

The effect is to situate the discourse in a hypothetical mood. Moreover, the language of the allusion itself is richly metaphorical, describing both the feeling of desire with an expression of divine love and the discernment of mental abstractions through the activity of beholding. Furthermore, Ptolemy explicitly draws the reader's attention to that latter metaphor, qualifying the verb 'to behold' with an adverbial 'as it were' (ὥσπερ).

This apology for the awkwardness of the visual metaphor is itself important, because it foregrounds the fact that Ptolemy does not simply adduce the madman-lover from the *Phaedrus*, but transforms it. Particularly in Socrates' second speech in the *Phaedrus*, Plato promotes the *spectacle* of beauty and the Forms, communicating his epistemology through metaphors of vision rather than other senses, such as hearing.[57] But the object that the theorist desires to behold is the 'nature of the harmonic power'—an abstraction perhaps more perceptible to the mind's ear. The upshot of Ptolemy's harmonic theory, then, is to reconcile reason and auditory perception: together they ascertain harmonic principles and the beauty manifest in them.[58] Through the allusion to the madman-lover of the *Phaedrus*, Ptolemy fashions his harmonic theory according to a Platonic model but refashions the model at the same time: consequent to Ptolemy's treatment, the beautiful has been made audible to the madman-lover.

Thus Ptolemy has creatively refashioned a memorable component of the *Phaedrus*, but what is most telling about the allusion is what it does not do. Nothing of Ptolemy's literary treatment of the madman-lover is necessary for the larger project of the *Harmonics*. As I noted at the outset, the allusion occurs in a transitional section of second-order discourse: it does not present any theoretical or technical content, nor does it explain such content, but serves only to fashion how the theorist in general (and Ptolemy and his reader, in particular) may feel about that content.[59] The allusion is thus extraneous to the overall presentation of harmonic theory.[60] Instead, it draws comparison to literary practices typical of second-century sophistic culture. Though authorial intentions ultimately lie beyond our grasp, we are left with a strong impression that in presenting the allusion Ptolemy is trying to be *interesting*.

[57]For the general importance of sight and spectacle to Platonic philosophy, especially as they relate to Plato's wider cultural context, see Nightingale 2004.

[58]Ptol. *Harm.* 1.2.1–31. Cf. Barker 2000: 14–32 and Barker 2010.

[59]In this the allusion is unique in the *Harmonics*. It is also true that Ptolemy adduces another, more general Platonic figure—the 'philosopher' (φιλόσοφος)—into his discussion at 3.5.70, but this is to illustrate further the concept of 'harmonia' (ἁρμονία); see Barker 1989: 377n50. In this passage Ptolemy does not use the philosopher to describe the practice of the theorist.

[60]In this it functions similarly to the 'frame tales' found in later mathematical commentaries that present famous mathematicians (e.g., Euclid) in moralizing episodes. These 'deliver not the [mathematical] knowledge itself, but rather the way a mathematician is supposed to behave when putting the knowledge to practical use' (Asper 2011: 96). A fundamental difference, however, is that Ptolemy is here not morally prescriptive, but emotionally so. He is idealizing the *experience*, not the behavior, of the harmonic theorist.

8 Mathematical Psychologies Prior to Ptolemy?

How comfortably does the expression of the desire for knowledge, Platonizing or otherwise, fit into the generic history of the exact sciences? The evidence surveyed thus far suggests that Ptolemy is unique in this regard: texts from harmonics, astronomy, and mathematics in general were markedly absent from our TLG surveys.[61] That search was restricted to certain defined phrases, of course, thus it will be worth examining the texts themselves to uncover any other expressions that eluded the TLG dragnet. However, any conclusions drawn from direct, textual examination must yet be qualified by the fact that so much evidence is either lost or fragmentary, but even the extant record reveals significant trends in authorial practice. Certain of these already suggest that the negative result of the TLG searches will be further confirmed. Reviel Netz, for instance, has observed an 'inflation of style' in Hellenistic mathematical writing, describing a gradual increase over time in the extent to which mathematical authors position and justify themselves *as authors* amidst a growing tradition of texts.[62] If Netz is generally correct, then we should not expect to find expressions of motivating desires among such earlier mathematical authors as Euclid, Autolycus, and Aristarchus, since such statements are often found in passages of authorial reflection. The second argument concerns the specifically Platonic character of Ptolemy's desire. It has often been observed that Hellenistic mathematical authors, for instance, focused almost exclusively on mathematics, and their writings typically betray little interest in other discursive practices.[63] When they do branch out, it tends to be into certain philosophical pursuits complementary to mathematics, such as astronomy.[64] By this account, Plato's psychology of the madman-lover might seem too far removed from actual mathematics to be of interest to mathematical writers. Taking the above into account, we thus offer two predictions: on the one hand, general claims of a desire for knowledge will only be found as a later development in exact-scientific writing (i.e., in Archimedes and after), if they are found at all; on the other hand, if such claims are made, they will be of a different character than Ptolemy's. The conclusion we shall arrive at generally bears out these predictions: especially in comparison with authors from the Hellenistic period, Ptolemy seems to have done something unique. What, then, do we find among his predecessors?

We begin not with the work of formal mathematicians, but with Ptolemy's early predecessors in harmonics, whom Ptolemy divides into opposing groups: the

---

[61]Nicomachus' *Introductio arithmetica* did register the expression σοφίας ὄρεξις (1.2.3), but this text is more a philosophical account of numbers than a presentation of geometric proofs (D'Ooge 1926: 16).

[62]Netz 2009: 92–107.

[63]Netz 1999: 306–311. Cf. Lloyd 1991: 369 and Taub 1993: 152.

[64]There are exceptions: below I examine Hipparchus' 'hybrid' commentary on Aratus. But we especially miss the lost works of the polymathic Eratosthenes, whose nonextant *Platonicus*, for instance, might have offered interesting counterevidence. On this work see Geus 2002: 141–194.

rationalist Pythagoreans and the empiricist Aristoxenians.[65] The Pythagorean harmonic tradition, exemplified by such figures as Philolaus of Croton or Archytas of Tarentum, survives only in fragments. Consequently, any analysis of self-expression in these authors is impeded by the bias of later writers who were more interested in, and thus more likely to record, first-order discussions of doctrine than any second-order statements on motivations. A survey of the fragments bears this out.[66] But we nevertheless do find testimony by Archytas that shows some reflection on harmonic practice: Archytas deems those concerned with 'mathematics' (μαθήματα, surely to be understood quite broadly[67]) to have 'discerned well' (καλῶς... διαγνῶναι) harmonic phenomena.[68] Note, however, that here Archytas' focus is how those practitioners operate ('well'), not what motivates them to do so. The motivations of those practitioners, as best as we can discern from our fragmentary evidence, are not preserved in the foreground of the text.

The work of Aristoxenus, the foremost proponent and namesake of the empiricist 'school' of harmonics, if not exactly a mathematician, has fared substantially better than that of the early Pythagoreans. Like the rationalists, however, Aristoxenus only reflects on the theorist insofar as method is concerned.[69] Regarding the psychological effects of the study of harmonics, Aristoxenus does note that some believe that 'listening to a discourse on harmonics (ἀκούσαντες τὰ ἁρμονικά) will make them not only experts in music (μουσικοί), but better in character (βελτίους τὸ ἦθος).'[70] But he is highly critical of this position, claiming that such individuals have misunderstood (παρακούσαντες) his statements concerning the limited effect that music itself may have on the hearer.[71] While Aristoxenus thus records a contemporary interest in the psychological effect that even the *theory* of harmonics may cause, he does not endorse it, nor does he specify any role that desire (or something like it) for harmonic theory might play. In any event, he does not promote an image of the harmonic theorist as one driven by a desire for knowledge. As far as our limited, fragmentary evidence indicates, earlier discourses on harmonics offered Ptolemy no positive, formal precedent for his presentation of the theorist.

Let us move on to texts that are more mathematical in form. As predicted above, numerous works in Hellenistic mathematics and mathematical astronomy offer no express characterization of the author or his motivations. These include the earliest of the genre (all those ascribed to Euclid, Autolycus, and Aristarchus), as well as later texts such as Hypsicles' *Anaphoricus* and the astronomical treatises of Theodosius. Their style is almost wholly impersonal, save for the conventional 'I say

---

[65] Ptol. *Harm.* 1.2.

[66] See Barker 1989: 28–52.

[67] On the connotations of 'mathematics' and 'mathematician' in antiquity, see Lloyd 2012.

[68] Reported in Porph. *In Harm.* 56.5 Düring.

[69] He is especially interested in defining the proper domain of the science of harmonics: see Aristox. *Harm.* 5.4–6.5 da Rios.

[70] Aristox. *Harm.* 40.14–16 da Rios, translation lightly adapted from Barker 1989: 148.

[71] Aristox. *Harm.* 40.16–41.2 da Rios.1 Cf. Barker 1989: 148n6.

that' (λέγω ὅτι), which is perhaps better understood as part of the formal structure of proof than as the interjection of an authentic, authorial self.[72] In all these the text privileges the presentation of mathematical research over the researcher, and thus there is little that they contribute to the discussion of expressions of desire.

Turning to Ptolemy's mathematical predecessors who evince a more personal style, we find no expression of desire in the terms that Ptolemy uses. Indeed, there are few explicit remarks about that of which desire is the corollary, namely beauty, but that beauty is not what we might expect. Consider how Apollonius of Perga describes certain theorems of the third book of his *Conics*: 'the third [contains] many paradoxical (παράδοξα) theorems . . . of which most and the most beautiful (κάλλιστα) are new' (vol. 1, p. 4 Heiberg). Apollonius does not make explicit what exactly defines those theorems as the 'most beautiful', and in general, Hellenistic mathematical texts do not offer much overt reflection on questions of aesthetics.[73] But consider that Apollonius' superlative, 'most beautiful', qualifies a subset of theorems primarily described as 'paradoxical'. Apollonius here seems to imply that the most beautiful theorems are those that are most paradoxical, thus implying a general aesthetic valuation of the unexpected (and the delight that it may prompt). Netz's recent work develops this notion that the Hellenistic mathematical aesthetic is characterized by surprise and variation, most evident in the narrative structures of mathematical texts, and which refract the 'Callimachean' aesthetic of contemporary Alexandrian poetics.[74] All of this amounts to a conception of beauty that, at least as expressed on the textual surface, is different from the metaphysical beauty exemplified by, say, Plato's Forms, and it is this Platonic conception—and the Platonic language—which Ptolemy adapts in his writings.

Given a beauty exemplified by a non-Platonic aesthetic of surprise and variation, it thus makes sense that mathematicians do not profess to react to it like Plato's eroticized philosophers. We find instead, simply, Hypsicles being 'charmed' (1.12 Stamatis: ἐψυχαγωγήθην)[75] by a problem of Apollonius, with no further comment. Further TLG searches indicate the term itself is a common feature of prose discourse from the Classical period onward; it does not evoke a particular author or text as did Ptolemy's expression. This leads to a general point about the Hellenistic mathematician's attitudes toward his subject: they are sometimes expressed in the text, but only in a passing, unmarked way. For instance, Archimedes, Eratosthenes, and Apollonius occasionally indicate their addressees' or predecessors' zeal for

[72]Netz 1999: 256.

[73]It is telling that a recent historical survey of mathematical aesthetics passes over Hellenistic mathematics entirely, leaping from Aristotle to Augustine (Sinclair and Pimm 2006: 4–5). On the beauty of mathematical objects in this context, see Netz 2005, esp. 282–283, and the next note.

[74]Netz 2009; cf. Netz 2005 and 2010. For similar studies of a similar aesthetic outside Greek mathematics, see Müller-Hill and Spies 2011: 266–268 and Schattschneider 2006.

[75]By Hypsicles' time, as in our own, a term whose semantics had stretched to include more figurative meanings.

mathematics using cognates of σπουδή, φιλοπονία, and φιλοτιμία.[76] They do not elaborate on these terms, however, nor is the language unique to the mathematical genre. As with cognates of ψυχαγωγία, TLG searches show that such expressions of zeal are common in Greek prose from the fourth century onward.[77] The key point is that Hellenistic mathematicians do register psychological motivations and attitudes in the text, but they *only just* register them. The motivations are presented in such a way that they do not stand out relative to other texts. For Hellenistic mathematicians, then, reactions to mathematics are not in themselves exciting. As Netz has shown, the excitement is to be unveiled—*voilà*!—in the mathematics. An enthusiastic reaction to them almost goes without saying.

Ptolemy is different. But the difference between Ptolemy's attitude and his forerunners' is not simply the expression of *desire*. This becomes clear upon examining the one extant work of Hipparchus, the second-century BCE astronomer whom Ptolemy elevates in the *Syntaxis* as an important predecessor. Hipparchus' *Commentary on the Phenomena of Aratus and Eudoxus* is, of course, several steps removed from mathematics: it is both a commentary on Aratus' hexameter poem, the *Phenomena*, which was based on Eudoxus' astronomical treatise of the same name, and a critique of another commentary on the same poem by one Attalus of Rhodes.[78] But given Hipparchus' importance to mathematical astronomy and to Ptolemy, in particular, it is an appropriate inclusion in the survey. And, as it turns out, an informative one: Hipparchus begins his treatise by addressing one Aischrion as follows: 'From your letter I gladly took note of your continuing propensity toward curiosity (φιλομαθίαν)' (Hipparch. 1.1.1). The theme of his addressee's curiosity or, more literally, his 'fondness for learning', is one that Hipparchus returns to again in the course of the preface, though he does not develop the idea beyond generalizing it, in participial form, to designate his wider readership.[79] The group of those who show commitment to astronomical theory is thus defined by the claim of a shared feeling ('fondness for learning'). This fondness thus functions in Hipparchus' text somewhat differently than zeal did in the mathematical writings surveyed above, in which the latter was always assigned exclusively to individuals. We may observe, on the other hand, that Hipparchus' fondness resembles Ptolemy's desire in two

[76]σπουδή: Archim. *Method* vol. 2, p. 71, col. 1.33–34 Netz et al.; Apollon.Perg. *Con.* vol. 1, p. 2 Heiberg. φιλοπονία: Archim. *Spir.* vol. 2, p. 2.18 Heiberg; Eratosthenes, at Eutoc. *In Archim. Sph. Cyl.* 90.4 Heiberg (authenticity defended by Knorr 1989: 131–146). φιλοτιμία: Apollon.Perg. *Con.* vol. 2, p. 2 Heiberg. In Toomer's translation of the Arabic copy of *On Burning Mirrors*, Diocles affirms that using gnomons requires 'care' (Toomer 1976: 42), which may have been σπουδή or something similar in the lost Greek original. The letter-form itself seems a natural vehicle to convey this attitude, since authors frequently offer their work expressly as the fulfillment of an eager correspondent's personal request for more mathematics.

[77]The same language even expresses the attitudes that euergetic Hellenistic kings and their subjugated polities show toward one another: see Ma 1999: 191.

[78]On the hybrid form of this work, see Netz 2009: 168–171.

[79]Aischrion's φιλομαθία: Hipparch. 1.1.5; 'those who are fond of learning' (τῶν φιλομαθούντων): Hipparch. 1.1.6, 1.10.25.

ways: first, with respect to the previous point, Ptolemy also described a feeling as the general condition of theorists in the exact sciences (including, we may infer, both himself and his reader); second, simply, Hipparchus' fondness and Ptolemy's desire convey the same basic idea.

Still, there is an important qualitative difference between φιλομαθία and Ptolemy's erotic desire. The former and its cognates, while they may have found their first flowering in the Platonic corpus, are widely found in other philosophical, rhetorical, geographical, medical, and historical authors from the fourth century BCE onward.[80] Thus φιλομαθία seems to be better classified as another species of the 'common' genus of the desire for knowledge. On the other hand, I have argued that Ptolemy's expression entails specifically Platonizing connotations, but it also proves an opportunity for reconfiguring a Platonic concept in a manner typical of second-century literary stylists. Hipparchus' fondness for learning does not really compare.

Ptolemy's desire stands without precedent in the exact sciences through the Hellenistic period. In the early Imperial period we can only grasp blindly for evidence: interest in mathematics did not vanish entirely,[81] but creative mathematical activity appears to have dried up.[82] One important witness survives, however, in Menelaus, an astronomer-mathematician of the late first century CE, apparently active in Rome, whose observations Ptolemy cites in the *Syntaxis*.[83]

Menelaus no longer speaks for himself, however: what remains of his work only survives in the Arabic tradition. Thanks to Abū Naṣr Manṣūr ibn 'Alī ibn 'Irāq, active around the turn of the first millennium,[84] we have essentially a complete translation of Menelaus' *Sphaerica*, a three-book treatise on spherical geometry. In the preface to this work, Menelaus appears to exemplify some of the literary practices described so far. Here I translate Krause's German text into English, referring as well to key phrases transliterated from the Arabic[85]:

> I know what lies in the proofs to make the soul receptive to them, and especially [the part] of those [proofs] in which there is beauty and to which belongs what the soul loves and desires (was die Seele liebt und begehrt: *wa kāna mimmā tuḥībbuhū al-nafsu wa-taštahīhi*). One can, if he loves learning (wenn er Belehrung liebt: *muḥibban li-t-ta'līmi*), make these things an instrument and build corollary theorems and problems out of them (117–18 Krause).

[80] The term appears frequently in the *Phaedo* and *Republic*, less so in the *Phaedrus*, *Timaeus*, and the *Laws*. Other Classical and Hellenistic instances include X. *An.* 1.9.5; Isoc. 1.18; Arist. *EN.* 1175a14; Plb. 1.2.8, etc.; Ps.-Scymnus 63; Aristeae.Ep. 1.6; Apollon.Cit. 3.15 Schöne. On its use especially in scientific texts, see Alexander 1993: 59, 100.

[81] This is the period of the philosophizing compiler Nicomachus: see n. 61 above. Cf. Cuomo 2000: 9–56 on the public profile of mathematics in the first centuries CE.

[82] This is the period Netz 1999: 284 describes as 'a wilderness between two deserts'.

[83] *Syntaxis* part 2, pp. 30 and 33 Heiberg; apparently the same Menelaus is present for the dialogue in Plutarch's *De facie* (930a).

[84] See further Krause 1998: 109–111.

[85] I am grateful to Alexander Key for his expertise and assistance on points of Arabic philology, which emerge in the following paragraphs.

Philological methods can only penetrate so far through layers of translation, but here I make a few observations. It seems clear that some form of the desire for knowledge underlies the phrase 'if he loves learning', and we can even offer a plausible reconstruction. The root of the verb *muḥibban*, *ḥbb*, typically denotes a generic, unmarked kind of love, consistent with the Greek root φιλ-; more emphatic expressions are often built on the root *š-h-y*, such as we find in the foregoing description of the soul's 'desire' (*wa-taštahīhi*). It seems a reasonable conjecture, then, that Menelaus' original expression featured some cognate of φιλομαθία. Indeed, we find that the Arabic translation of Aristotle's *Nicomachean Ethics* 1175a14, roughly contemporary with Abū Naṣr,[86] renders ὁ φιλομαθής as *al-muḥibb fī 't-ta'līm*, [87] a phrase almost identical to Abū Naṣr's expression.[88] The difference seems due simply to the definite article, present in Aristotle's text and plausibly lacking from Menelaus'. For the latter, it is easy to imagine either a participial or adjectival cognate of φιλομαθία, perhaps in the dative case as part of an impersonal construction. It seems clear enough, then, that Menelaus' theorist of spherical geometry should feel a desire for knowledge—but was it a Platonizing desire?

This is harder to ascertain, but some plausibility lies in the expression of '[the part] of those [proofs] in which there is beauty and to which belongs what the soul loves and desires (was die Seele liebt und begehrt: *wa kāna mimmā tuḥībbuhū al-nafsu wa-taštahīhi*)'. Krause noted that the account of beauty and the loving and desiring soul features Quranic overtones.[89] This probably does not indicate an interpolation by Abū Naṣr, but it could be an instance of an artful and learned translator seizing an opportunity to draw his source material closer to a work of high cultural value. But the very same elements, if found in the Greek, would fit well in a Platonizing expression, and the language calls to mind the *Phaedrus*. The first-century CE Menelaus is active within the period when we expect this to occur, and if the expression 'loves and desires' faithfully transmits a duplex structure in Menelaus' original, it is again plausible that one of those terms might have been a cognate of ἔρως (though other Greek terms of 'desire' could just as well have comprised the original). It is at least a possibility, then, that in the *Sphaerica*, Menelaus featured a Platonizing expression for the desire for knowledge. Ptolemy is perhaps then neither the only nor the first author in the exact sciences to exemplify such passion. Lacking Menelaus' Greek, we cannot determine whether he likewise might have anticipated Ptolemy in creatively refashioning his source material. Even without that final flourish, however, the evidence of Menelaus encourages us to speculate that in the Imperial period, and in contrast to generic

[86]The extant manuscript is itself dated 1222 CE, but the translation apparently dates to the ninth or tenth century: see Akasoy and Fidora 2005: 1–2.

[87]Akasoy and Fidora 2005: 544n91.

[88]Does this indicate a convention shared among translators for rendering φιλομαθ- into Arabic?

[89]Krause 1998: 117n5.

precedents, the mathematical sciences kept pace with contemporary prose genres by conceptualizing theoretical motivations according to a Platonizing aesthetic.

9 Ptolemy's Timely Desires

In drawing this paper to a close, let us consider what insights we have gained through our examination of Ptolemy's desire for knowledge. That the desire for knowledge is fundamental to human nature? This much was assumed at the outset. Heisenberg could have been just as certain that Ptolemy, like Einstein, must have felt something that moved him toward mathematical investigations, and I can make no claim to revealing anything new about Ptolemy's psychology unless I should first prove or disprove his humanity. But what was the 'something' that Ptolemy felt? There should be no surprise at the basic fact that expressions of the desire for knowledge are shaped by culture; in this it is a desire like any other. But I hope to have shown in this essay that the shapes themselves are significant and invite philological and historical analysis. Heisenberg confesses his 'attraction': this is the expression of desire as an impersonal force at perhaps its utmost demystification, a mere 'drawing towards', even when its object is beauty. Is it a sign of Heisenberg's times? He is certainly no second-century Platonizer 'in the grip of a divine Eros'. In the expression of psychological motivations, then, we find one salient difference (among many) between Heisenberg and Ptolemy. I have argued that similar differences, traced along generic and chronological lines, prevailed in Greek literary culture from the Classical period through the Imperial, and have explicated how some of these differences were significant. Focusing on Ptolemy, I examined the manifestation and significance of the Platonizing expression of the desire for knowledge in Imperial literary culture, and how the forms and processes of creative reconfiguration it entails serve as points of continuity between exact-scientific works and texts in other prose genres. In doing so, I promoted a view of Ptolemy, in particular, as something of a prose-stylist. This paper began by considering Ptolemy's apparent isolation from his wider culture, and now it closes by qualifying that isolation: Ptolemy's activity in a specialized genre may have been directed toward and received by a specialized readership, but the motivations he expresses for engaging in that activity and the creative manner in which he expresses them are nevertheless founded in the wider literary culture of his time.

References

Akasoy, A. A., & Fidora, A. (Eds.). (2005). *The Arabic version of the* Nicomachean ethics. *With an introduction and annotated translation by D. M. Dunlop*. Leiden and Boston: Brill.

Alexander, L. (1993). *The preface to Luke's Gospel: literary convention and social context in Luke 1.1–4 and Acts 1.1*. Cambridge: Cambridge University.

Arrowsmith, W. (1973). Aristophanes' *Birds:* the fantasy politics of *eros*. *Arion,* 1, 119–167.

Asper, M. (2007). *Griechische Wissenschaftstexte. Formen, Funktionen, Differenzierungsgeschichten*. Stuttgart: Steiner.

Asper, M. (2011). "Frame tales" in ancient Greek science writing. In K.-H. Pohl & G. Wöhrle (Eds.), *Form und Gehalt in Texten der Griechischen und Chinesischen Philosophie* (pp. 91–112). Stuttgart: Steiner.

Barker, A. (Ed.). (1989). *Greek musical writings, Vol. II: Harmonic and acoustic theory*. Cambridge: Cambridge University Press.

Barker, A. (2000). *Scientific method in Ptolemy's harmonics*. Cambridge: Cambridge University Press.

Barker, A. (2010). Mathematical beauty made audible: musical aesthetics in Ptolemy's harmonics. *Classical Philology,* 105, 403–420.

Barton, T. S. (1994). *Power and knowledge: astrology, physiognomics, and medicine under the Roman empire*. Ann Arbor: University of Michigan Press.

Bernard, A. (2010). The significance of Ptolemy's *Almagest* for its early readers'. *Revue de Synthèse,* 131, 495–521.

Connor, W. R. (1992). *The new politicians of fifth-century Athens*. Indianapolis: Hackett [First published (1971). Princeton: Princeton University Press].

Cornford, F. M. (1907). *Thucydides mythistoricus*. London: Edward Arnold.

Cuomo, S. (2000). *Pappus of Alexandria and the mathematics of late antiquity*. Cambridge: Cambridge University Press.

da Rios, R. (Ed.). (1954). *Aristoxeni elementa harmonica*. Rome: Officina Poligrafica.

De Lacy, P. (1972). Galen's Platonism. *American Journal of Philology,* 93, 27–39.

De Lacy, P. (1974). Plato and the intellectual life of the second century A.D. In G. W. Bowersock (Ed.), *Approaches to the Second Sophistic* (pp. 4–10). University Park, PA: American Philological Association.

Dillon, J. (1977). *The Middle Platonists: a study of Platonism, 80 B.C. to A.D. 220*. London: Duckworth.

D'Ooge, M. L. (Ed.). (1926). *Nicomachus of Gerasa: introduction to arithmetic. Translated with studies in Greek arithmetic by F. E. Robbins & L. C. Karpinski*. New York: Macmillan.

Düring, I. (Ed.). (1930). *Die Harmonielehre des Klaudios Ptolemaios, Göteborgs Högskolas Arsskrift 36*. Gothenburg: Elanders Boktryckeri Aktiebolag.

Düring, I. (Ed.). (1932). *Porphyrios. Kommentar zur Harmonielehre des Ptolemaios*. Gothenburg: Elanders Boktryckeri Aktibolag.

Feke, J. (2009). *Ptolemy in philosophical context: a study of the relationships between physics, mathematics, and theology*. Ph.D. thesis, University of Toronto, Toronto.

Feke, J. (2012). Ptolemy's defense of theoretical philosophy. *Apeiron,* 45, 61–90.

Feke, J., & Jones, A. (2010). Ptolemy. In L. Gerson (Ed.), *Cambridge history of philosophy in late antiquity* (Vol. 1, pp. 197–209). Cambridge: Cambridge University Press.

Fögen, T. (2009). *Wissen, Kommunikation und Selbstdarstellung: zur Struktur und Charakteristik römischer Fachtexte der frühen Kaiserzeit*. Munich: Beck.

Fuhrmann, M. (1960). *Das systematische Lehrbuch. Ein Beitrag zur Geschichte der Wissenschaft in der Antike*. Göttingen: Vandenhoeck & Ruprecht.

Geus, K. (2002). *Eratosthenes von Kyrene: Studien zur hellenistischen Kultur- und Wissenschaftsgeschichte*. Münchener Beiträge zur Papyrusforschung und antiken Rechtsgeschichte 92. Munich.

Gill, C., Whitmarsh, T., & Wilkins, J. (Eds.). (2009). *Galen and the world of knowledge*. Cambridge: Cambridge University Press.

Halperin, D. (1985). Platonic *erôs* and what men call love. *Ancient Philosophy,* 5, 161–204.

Heiberg, J. L. (Ed.). (1891–1893). *Apollonii Pergaei quae Graece exstant* (Vol. 2). Leipzig: Teubner.

Heiberg, J. L. (Ed.). (1898–1903). *Claudii Ptolemaei Opera quae exstant omnia, Vol. 1, Syntaxis Mathematica (2 parts)*. Leipzig: Teubner.

Heiberg, J. L. (Ed.). (1910–1915). *Archimedis Opera omnia cum commentariis Eutocii* (3 vols.). Leipzig: Teubner.

Heisenberg, W. (1971). *Physics and beyond: encounters and conversations. Translated from the German by J. Arnold*. New York: Pomerans.

Hinds, S. (1998). *Allusion and intertext: dynamics of appropriation in Roman poetry*. Cambridge: Cambridge University Press.
Hunter, R. (1983). *A study of Daphnis and Chloe*. Cambridge: Cambridge University Press.
Hunter, R. (1997). Longus and Plato. In M. Picone & B. Zimmermann (Eds.), *Der antike Roman und seine mittelalterliche Rezeption* (pp. 15–28). Basel: Birkhäuser Verlag.
Hunter, R. (2012). *Plato and the traditions of ancient literature: the silent stream*. Cambridge: Cambridge University Press.
Jones, A. (1990). *Ptolemy's first commentator*. Transactions of the American Philosophical Society 80.7. Philadelphia.
Jones, A. (2004). An "*Almagest*" before Ptolemy's? In C. Burnett, J. P. Hogendijk, K. Plofker, & M. Yano (Eds.), *Studies in the history of the exact science in honour of David Pingree* (pp. 129–136). Leiden: Brill.
Jones, A. (2005a). Ptolemy's "Canobic inscription" and Heliodorus's observation reports. *SCI-AMVS: Sources and Commentaries in Exact Sciences*, 6, 53–97.
Jones, A. (2005b). In order that we should not ourselves appear to be adjusting our estimates . . . to make them fit some predetermined amount. In J. Z. Buchwald & A. Frankl (Eds.), *Wrong for the right reasons* (pp. 17–39). Dordrecht: Springer.
Jones, A. (2010). Introduction. In A. Jones (Ed.), *Ptolemy in perspective: use and criticism of his work from antiquity to the nineteenth century* (pp. 11–15). Dordrecht: Springer.
Knorr, W. R. (1989). *Textual studies in ancient and medieval geometry*. Boston: Birkhauser.
Krause, M. (1998). *Die Sphärik von Menelaos aus Alexandrien in der Verbesserung von Abû Nasr Mansûr B. 'Alî B. 'Irâq, mit Untersuchungen zur Geschichte des Textes bei den Islamischen Mathematikern*. Frankfurt am Main: Institute for the History of Arab-Islamic Science. [First published (1936). Berlin: Weidmann.]
Leigh, M. (2013). *From* polypragmon *to* curiosus*: ancient concepts of curious and meddlesome behaviour*. Oxford: Oxford University Press.
Lloyd, G. E. R. (1991). Science and morality in Greco-Roman antiquity. In G. E. R. Lloyd (Ed.), *Methods and problems in Greek science* (pp. 352–371). Cambridge: Cambridge University Press. [First published as (1985) *Science and morality in Greco-Roman antiquity: an inaugural lecture*. Cambridge: Cambridge University Press.]
Lloyd, G. E. R. (2012). The pluralism of Greek "mathematics". In K. Chemla (Ed.), *The history of mathematical proof in ancient traditions* (pp. 294–310). Cambridge: Cambridge University Press.
Ma, J. (1999). *Antiochus III and the cities of western Asia Minor*. Oxford: Oxford University Press.
Mansfeld, J. (1994). *Prolegomena: questions to be settled before the study of an author or a text*. Leiden: Brill.
Mansfeld, J. (1998). *Prolegomena mathematica: from Apollonius of Perga to late Neoplatonism, with an appendix on Pappus and the history of Platonism*. Leiden: Brill.
Mattern, S. (2008). *Galen and the rhetoric of healing*. Baltimore: Johns Hopkins University Press.
Müller-Hill, E., & Spies, S. (2011). Der Begriff mathematischer Schönheit in einer empirisch informierten Ästhetik der Mathematik. In M. Helmerich, K. Lengnink, G. Nickel, & M. Rathgeb (Eds.), *Mathematik Verstehen: Philosophische und Didaktische Perspektiven* (pp. 261–281). Wiesbaden: Vieweg & Teubner Verlag.
Netz, R. (1997). Classical mathematics in the classical Mediterranean. *Mediterranean Historical Review*, 12(2), 1–24.
Netz, R. (1999). *The shaping of deduction in Greek mathematics: a study in cognitive history*. Cambridge: Cambridge University Press.
Netz, R. (2002). Greek mathematicians: a group picture. In L. Wolpert, C. J. Tuplin, & T. E. Rihll (Eds.), *Science and mathematics in ancient Greek culture* (pp. 196–216). Oxford: Oxford University Press.
Netz, R. (2005). The aesthetics of mathematics: a study. In P. Mancosu, K. F. Jørgensen, & S. A. Pedersen (Eds.), *Visualization, explanation and reasoning styles in mathematics* (pp. 251–293). Dordrecht: Springer.

Netz, R. (2009). *Ludic proof: Greek mathematics and the Alexandrian aesthetic*. Cambridge: Cambridge University Press.
Netz, R. (2010). What did Greek mathematicians find beautiful? *Classical Philology*, 104, 426–444.
Netz, R., Noel, W., Tchernetska, N., & Wilson, N. (Eds.). (2011). *The Archimedes palimpsest (2 vols.)*. Cambridge: Cambridge University Press.
Nightingale, A. W. (1996). *Genres in dialogue: Plato and the construct of philosophy*. Cambridge: Cambridge University Press.
Nightingale, A. W. (2004). *Spectacles of truth in classical Greek philosophy:* theoria *in its cultural context*. Cambridge: Cambridge University Press.
Reardon, B. P. (Ed.). (1989). *Collected ancient Greek novels*. Berkeley: University of California Press.
Rocca, J. (2006). "Plato will tell you": Galen's use of the *Phaedrus* in *De Placitis Hippocratis et Platonis* IX. In H. Tarrant & D. Baltzly (Eds.), *Reading Plato in antiquity* (pp. 49–59). London: Duckworth.
Rota, G.-C. (1997). The phenomenology of mathematical beauty. *Synthese*, 111, 171–182.
Rothwell, K. S., Jr. (1990). *Politics and persuasion in Aristophanes'* Ecclesiazusae. Leiden: Brill.
Schattschneider, D. (2006). Beauty and truth in mathematics. In N. Sinclair, D. Pimm, & W. Higginson (Eds.), *Mathematics and the aesthetic: new approaches to an ancient affinity* (pp. 42–57). New York: Springer.
Sinclair, N., & Pimm, D. (2006). A historical gaze at the mathematical aesthetic. In N. Sinclair, D. Pimm, & W. Higginson (Eds.), *Mathematics and the aesthetic: new approaches to an ancient affinity* (pp. 1–19). New York: Springer.
Stamatis, E. S. (ed. post J. L. Heiberg). (1977). *Euclidis elementa, Vol. 5.1, Elementa XIV, XV, Scholia in Libros I–V.* Leipzig: Teubner
Swerdlow, N. M. (2004). Ptolemy's harmonics and the "tones of the universe" in the *Canobic inscription*. In C. Burnett, J. P. Hogendijk, K. Plofker, & M. Yano (Eds.), *Studies in the history of the exact sciences in honour of David Pingree* (pp. 137–180). Leiden: Brill.
Tarrant, H. (2000). *Plato's first interpreters*. London: Duckworth.
Taub, L. C. (1993). *Ptolemy's universe: the natural philosophical and ethical foundations of Ptolemy's astronomy*. Chicago: Open Court Press.
Tihon, A. (2010). An unpublished astronomical papyrus contemporary with Ptolemy. In A. Jones (Ed.), *Ptolemy in perspective: use and criticism of his work from antiquity to the nineteenth century* (pp. 1–10). Dordrecht: Springer.
Tolsa, C. (2013). *Claudius Ptolemy and self-promotion: a study on Ptolemy's intellectual milieu in Roman Alexandria*. Ph.D. thesis, University of Barcelona, Barcelona.
Toomer, G. J. (1975). Ptolemy. In C. C. Gillespie (Ed.), *Dictionary of scientific biography (18 vols.)* (Vol. 11, pp. 186–206). New York: Charles Scribner's Sons.
Toomer, G. J. (Ed.). (1976). *Diocles, On burning mirrors: the Arabic translation of the lost Greek original. With English translation and commentary*. Berlin: Springer.
Toomer, G. J. (1985). Galen on the astronomers and astrologers. *Archive for the History of the Exact Sciences*, 32, 193–206.
Trapp, M. B. (1990). Plato's *Phaedrus* in second-century Greek literature. In D. A. Russell (Ed.), *Antonine literature* (pp. 141–174). Oxford: Oxford University Press.
Trapp, M. B. (Ed. and Trans.). (1997). *Maximus of Tyre: the philosophical orations*. Oxford: Clarendon.
van der Eijk, P. J. (1997). Towards a rhetoric of scientific discourse: some formal characteristics of Greek medical and philosophical texts. In E. J. Bakker (Ed.), *Grammar as interpretation: Greek literature in its linguistic contexts* (pp. 77–129). Leiden: Brill.
Whitmarsh, T. (2001). *Greek literature and the Roman empire: the politics of imitation*. Oxford: Oxford University Press.
Whitmarsh, T. (2005). *The Second Sophistic*. Oxford: Oxford University Press.

Transmission and Interactions Among Different Types of Geometrical Argumentations: From Jesuits in China to Nam Pyŏng-Gil in Korea

Jia-Ming Ying

Abstract

Transmission and interactions among different types of geometrical argumentations constitute some of the most interesting stories in the history of East Asian science and mathematics. Since the early years of the seventeenth century, the Chinese began to learn European science and mathematics introduced by missionaries and incorporate these into translations and into their own works. At first, Euclid's *Elements*, with its hypothetico-deductive structure, was translated in early 1600s. However, one of the most influential mathematical treatises in late imperial China and in contemporary Korea, the *Shuli jingyun* (Essential Principles of Mathematics, 1723), was composed as a synthesis of all the Chinese and European mathematical knowledge that was available to the Qing emperor Kangxi (r. 1662–1722) himself and his royal mathematicians. A section in this mathematical compendium is entitled "*Jihe yuanben*" (Elements of geometry), which does not refer to the first Chinese translation of Euclid's *Elements* bearing the same Chinese title. It is actually taken from lecture notes written by the French Jesuits Jean-François Gerbillon and Joachim Bouvet when they taught mathematics to Kangxi in the 1690s. These notes were in turn based on the

In this chapter, the McCune-Reischauer system and the Pinyin system are used for Romanising all Korean and Chinese words, respectively, except for the English articles in which authors' names and titles have been Romanised in other ways. Koreans also use Chinese characters, so the Korean names, book titles or terminologies that are written in Chinese characters but originally used by Koreans are Romanised according to the McCune-Reischauer system for the Korean language.

J.-M. Ying (✉)
College of Humanities and Social Sciences, Taipei Medical University, No. 250, Wu-Hsing Street, Xinyi Distrct, Taipei, 110 Taiwan
e-mail: j.m.ying@tmu.edu.tw

S. Ju et al. (eds.), *Cultures of Mathematics and Logic*, Trends in the History of Science, DOI 10.1007/978-3-319-31502-7_6

French geometry textbook *Elémens de Géométrie* by the Jesuit Ignace-Gaston Pardies.

The style of argumentation in Pardies' text is to give quick and easy explanations that appeal not entirely to the rigour of logic but to the intuition of the reader, and this pedagogy was used by Gerbillon and Bouvet in their lecture notes, which was later compiled into the *Shuli jingyun*. Some cases can be found to show this style of argumentation. For the volume of pyramids, basically the argument is that a cube could be cut into three "pointed solids", so the latter's volume was one third of a cube, and then the volumes of all other pyramids and cones could be calculated by the same procedure, because pointed solids with equal base areas and equal heights would have equal volumes. For the relation between the surface area and the volume of the sphere, the reader is asked to imagine that the sphere is composed of "millions of tiny cones" whose bases are parts of the surface of the sphere, and whose heights are equal to the radius of the sphere.

The *Shuli jingyun* was transmitted to Korea shortly after its publication, and its influence can be seen on many cases, including the arguments on the volume of different kinds of pyramids and that of the sphere written in the Korean commentary for the *Jiuzhang suanshu* (Nine Chapters of Mathematical Art). The Korean commentator Nam Pyŏng-Gil intentionally replaced the traditional Chinese commentary on the two problems with his explanations that appeal mainly to intuition. This very style and its transmission is an interesting example of how mathematicians in pre-modern China and Korea chose their ways of composing texts and arguing mathematical propositions.

1 Introduction

The first Jesuit to enter China, Matteo Ricci (1552–1610), arrived at Macao in early 1580s. He has come down in history as the "founding father" of the Jesuit mission in China.[1] For the first two centuries of the Jesuits' presence in China, since 1582, they put their science in the service of the evangelisation: the knowledge and know-how that they displayed enhanced the prestige of Christianity and served to attract the patron of some officials and that of the imperial state (Jami 2012, p. 13). They had first presented science as forming a coherent whole with the Catholic religion, a whole that was sometimes referred to as "heavenly learning" (*tianxue* 天學), which included various topics taught in Jesuit colleges as well as the Catholic religion itself. However, what had been received by Chinese scholars, known as the "Western leaning" (*xixue* 西學), comprised mainly astronomy and mathematics (Jami and Han 2003). In the seventeenth and eighteenth centuries, many texts related to "Western" mathematics written in Chinese or Manchu by Jesuit missionaries or Chinese scholars, were composed and often published. Those texts had a significant

[1] See, for instance, Spence (1984).

impact on the Chinese literati who were versed in the subject. Some of those texts were also brought to the possessions of scholars in other parts of East Asia.

For some of China's neighbouring countries, such as Korea, the transmission of Western learning was mainly based on Chinese sources. Unlike Japan, into which the Jesuits were allowed to enter in the sixteenth and seventeenth centuries, Korea's Chosŏn 朝鮮 Dynasty (1392–1910) before the nineteenth century never willingly allowed Westerners to enter their country.[2] The Chinese sources about Western learning also shaped the development of Korean mathematics from the seventeenth to the nineteenth centuries.[3] Two interesting questions can be asked in this context. First, what parts of European mathematics were included in these Chinese texts and later studied and used by Koreans? And more importantly, what styles of problem solving and mathematical argumentations can be found in Korean texts after these Chinese texts containing Western knowledge had been acquired?

The topic on the styles of argumentations is one of the particularly interesting issues in East Asian mathematics. In 1607, the first six books of Christopher Clavius' edition of Euclid's *Elements*, translated by Ricci and Xu Guangqi 徐光啓 (1562–1633) with the title *Jihe yuanben* 幾何原本 (Elements of Mathematics), was published in Beijing.[4] Ricci and Xu's translation was highly praised by many Ming and Qing mathematicians.[5] However, there is another mathematical text that was also influential in the Qing period. In the eighteenth and nineteenth centuries, one of the most influential mathematical texts in China was the *Yuzhi Shuli jingyun* 御製數理精蘊 (Essential Principles of Mathematics, Imperially Composed, 1723), from here on abbreviated as the *Shuli jingyun*. It also includes a substantial section whose title, *Jihe yuanben*, is exactly the same as that of Ricci and Xu's translation (Guo 1993). At first glance, this version of the "Elements" also talks about Euclidean geometry, but its styles of writing and composition are very different from those in Ricci and Xu's translation of Euclid's *Elements*. Some research questions naturally arise from this observation: What are the sources for the compilation of this *Jihe yuanben*? Logically speaking, how are its styles of argumentation different from that of Euclid's *Elements*? And what are its influences on the styles of argumentation by Korean mathematicians, who depended on Chinese sources for the study of

[2]For the transmission of astronomical and mathematical knowledge into Japan by the Jesuits, see, for instance, Horiuchi (2010), pp. 6–7.

[3]The reader may refer to, for example, Guo (2009), pp. 38–46, 74–75.

[4]See Engelfriet (1998). The term "*jihe*" 幾何 in classical Chinese does not originally mean "geometry" as the term is meant in modern Mandarin. In classical Chinese the term simply means "how many/much". In ancient Chinese mathematical texts, a question usually ends with this term "*jihe*", so Ricci and Xu used it to represent "mathematics". However, when the French Jesuits compiled their lecture notes according to Pardies' *Elémens de Géométrie* (see below) in late seventeenth century, they also named it the *Jihe yuanben*. So for this later text, *jihe* is indeed used to mean "geometry". See Tian (2003), pp. 26–38.

[5]For the influence of Ricci and Xu's translation of Euclid's *Elements*, refer to, for instance, Mei et al. (1990), pp. 53–83.

Western learning? In this article, I will discuss the sources and give some examples of geometrical propositions to give partial answers to these questions.

2 Sources and the Style of Argumentations of "Elements of Geometry" in Emperor Kangxi's *Shuli jingyun*

As mentioned before, Ricci and Xu's translation of the first six books of Euclid's *Elements* was published in 1607. Generally speaking, the style of argumentation in Ricci and Xu's translation, as in Clavius edition and in the original Greek text, is to give justifications to the legitimacy of constructions, or observed relations among components, of a diagram *within a system*, the conventions and rules of which have been stated in the definitions, common notions and postulates at the beginning of each book (Jami 1996, p. 190). However, this text and its styles of argumentation were not included in emperor Kangxi's 康熙 (r. 1662–1722) mathematical compendium.

Kangxi's mathematical compendium, the *Shuli jingyun*, is the mathematical section of the *Lüli yuanyuan* 律曆淵源 (Source of Pitch-pipes and the Calendar), an even greater compendium of mathematics, astronomy and musical harmony. The creation of the *Lüli yuanyuan* was part of Kangxi's policy of appropriation of Western knowledge. During the early years of Kangxi's reign, Yang Gunagxian 楊光先 (1597–1669) lead attacks on the Jesuits who worked in the Astronomical Bureau (*Qintianjian* 欽天監) on the grounds of both astronomical calculations and divination, which took away the Jesuits' favours among the young emperor's Manchu regents, and lead to the 1664 impeachment. This is the famous "Calendar Case" (*Li yu* 曆獄). The verdicts of this case were later revoked in 1669, partly due to the erroneous calculations given by Yang and his team after they took over the Astronomical Bureau, but mainly thanks to the Jesuit Ferdinand Verbiest (1623–1688) with his correct astronomical observations and predictions. Kangxi himself was shocked by the incompetence of his Chinese astronomers, who failed to help the emperor of his imperial duty of granting a correct calendar and of securing the harmony between the heavens and human life. And that could have vital threats to the new empire which had ruled China proper for only a few decades.[6] The calendar case was one of the reasons that prompted Kangxi in his own studies of Western learning, and his wishes to compile that knowledge into a new compendium for the foundation of his empire (see Han 1996).

In 1713, a group of young scholars versed in mathematics was brought together in the newly founded Office of Mathematics (*Suanxue guan* 算學館) under the imperial patronage of Kangxi. The reason to establish such a place was Kangxi's desire to have an institution independent of the Astronomical Bureau in which the Jesuits worked. On the one hand Kangxi believed that the study of astronomy should not be mixed with the Europeans' mission to convert the subject of his empire, and on the

[6]For this calendar case, the reader may refer to Huang (1991). Or refer to Chu (1997).

other hand he wanted the empire to have a firm foundation of scientific knowledge that was not dependent on the skills of foreigners. One major purpose of the Office of Mathematics was to compile the *Lüli yuanyuan*.[7] The Office was supervised by Prince Yinzhi 胤祉 (1677–1732), Kangxi's third son, and the mathematicians He Guozong 何國宗 (late-17th c–1766) and Mei Juecheng 梅瑴成 (1681–1763, a grandson of Mei Wending's) were the main editors of the whole compendium.[8] Eventually the *Shuli jingyun* was composed as a synthesis of all the European and Chinese mathematical knowledge that was available to Kangxi himself and to his court mathematicians.[9] Soon after its publication in 1723, the *Shuli jingyun* became the most important source for Chinese scholars who were interested in Western mathematics. As a matter of fact, it remained a compulsory text for mathematical instruction until the middle of the nineteenth century (Horng 1993).

The *Shuli jingyun* is divided into First Half (*Shang bian* 上編) and Second Half (*Xia bian* 下編). The second to the fourth volume (*juan* 卷) of the First Half is the *Jihe yuanben* (Guo 1993, pp. 20–141). Many studies have shown that the contents of this *Jihe yuanben* mainly came from the lecture notes written by the French Jesuits Jean-François Gerbillon (1654–1707) and Joachim Bouvet (1656–1730) when they taught mathematics to Kangxi as his court savants in the 1690s. These notes were in turn based on the French geometry textbook *Elémens de Géométrie* by the Jesuit Ignace-Gaston Pardies (1636–1673).[10]

If one tries to compare the structures of Pardies' *Elémens*, Gerbillon and Bouvet's lecture notes, and the *Jihe yuanben* in the *Shuli jingyun*, then one would find that the majority of the mathematical contents are identical. All topics, except for the progressions and logarithms in Book Eight of the *Elémens*, were translated into the Jesuits' lecture notes and later compiled into Kangxi's compendium (Liu 1991, pp. 88–96). Pardies' intention with the text was clear from its full title: "Elements of Geometry, in which by a short and easy method one can learn all one should know about Euclid, Archimedes, Apollonius, and the most beautiful inventions of ancient and new geometers". And he points out in his preface the difficulties in studying Euclid:

> Maybe, after all, people will think I write things down only in an abbreviated form, & that this Geometry can serve as a memorandum to those who already know this science, but not as an instruction for those who want to study it. I say that this is very remote from my intention, which was never to make an epitome; I always intended to make a Geometry that could be used to those who begin, and in which even those who have never heard of Mathematics could learn in very little time not only what is most necessary in Geometry, but also what is most elevated [...] For it should be noted that one of the things that make the reading of Euclid and of ordinary Authors difficult and boring, is that in the rigorous exactness that they have put to let nothing that can be proved pass without proof, easy as it may seem otherwise, it often happens that what would have been clear if one had

[7]Han (1999). Also refer to Han (2008), pp. 967–986.

[8]Jami and Han (2003). And refer to Han (2003).

[9]See Han (1995), pp. 87–127. Or refer to Jami (2004), pp. 92–93.

[10]See, for instance, Han (1993), pp. 1–10; or Jami (1996).

> been content with proposing it to the mind, as it naturally appears, becomes difficult and embarrassed, when one wants to reduce it to demonstration. [11]

Indeed, if one reads this text, one would find that Pardies does not differentiate definitions from propositions, and most statements were simply given without any justification, because "they naturally appeared clear" (Pardies 1673).

Pardies' pedagogy can be categorised as an "intuitive" approach in mathematics education. The notion "intuition" is usually used in different ways in the philosophy of mathematics community and by mathematicians or mathematics educators. The notion of intuition is certainly a topic of debates in the philosophy of mathematics community. One of the most intriguing and controversial suggestions in philosophy of mathematics, raised by the famous logician Kurt Gödel, is his realism in the sense that mathematical objects, such as numbers or objects of set theory, may exist independently of our constructions, that we have "intuition" (or something like a "perception") of these objects, and that the axioms of those objects "force themselves on us as being true". Thus for Gödel, our "intuition" of mathematical objects is the entrenched beliefs about them.[12] However, in this article the term "intuition" is not used in the Gödelian sense, but in the same sense as many mathematicians and mathematics educators use it. For them the term usually means a revelation or understanding about a mathematical proposition, achieved through mental manipulation about certain mathematical objects, or simple observation of some physical representations of those objects.[13] When a person has an "intuition" about some proposition, it does not guarantee that she know its proof in the strict sense. It does mean that the person has some idea about why the proposition is true, or false, in a system. We shall see several examples in this paper about how an "intuitive" approach of mathematical arguments is done in Pardies' and some mathematicians' works in East Asia.

Pardies' pedagogy of putting emphasis on intuition over the method of synthesis was not unique in the seventeenth century. In fact, his *Elémens* was but one of dozens of textbooks produced in Europe in that century that employed intuition, numerical thinking and the analytic method instead of adhering to Euclid's ways of organising knowledge in geometry (Karpinski and Kokomoor 1928; Kokomoor 1928).

The French Jesuits Gerbillon and Bouvet made good use in their lecture notes of that seventeenth-century trend in pedagogy of geometry and criticism on Euclid's

[11]The English translation of Pardies' title and preface is quoted from Jami (1996), pp. 187–188.

[12]For a general overview of Gödel's notion of intuition, the reader can refer to Shapiro (2000), pp. 202–211. A noteworthy discussion is in Parsons (1979), pp. 144–168.

[13]About how mathematicians describe their "intuition", the reader can refer to Burton (1999).

Elements.[14] In fact, the preface to the Chinese edition of their lecture notes begins as follows:

> 幾何原本
> 數源之謂,利馬竇所著.因文法不明,後先難解,故另譯.[15]
> 乃度數萬物之根本[...]由得圖形而明情理,不求講解而得其詳論也.[16]
>
> The *Elements of Mathematics*
> [It] means the origins of [the methods of] numbers. It was written by Matteo Ricci. Since it is unclear in its composition and it is difficult to elucidate what comes later from what comes first, [we have made] another translation.
> is the fundamentals for understanding numbers and a myriad of things. [...] From the diagrams [we] have, [we can] elucidate the conditions and reasons. [We] do not seek to explain in words but [we may] obtain detailed arguments.

Thus the style of argumentation in the lecture notes, and subsequently in the *Shuli jingyun*, appeals more to intuition than to rigour. In what follows, two groups of examples, one about the volumes of pyramids and cones and the other about the relation between the volume and surface area of a sphere, will be used to discuss the style of argumentations in the Jesuits' texts, in the *Shuli jingyun*, and also in a Korean source that will be mentioned in the next section.

Articles 32 and 33 in Book Five of Pardies' *Elémens* give the following statements without proof: "Pyramids and cones on equal bases, and contained in the same parallels, with prisms and cylinders, are one third of such prisms and cylinders"; "Every sphere is equal to a cone whose perpendicular axis is the radius of the sphere, and its base a plane, equal to all the convex surface of it" (Pardies 1673, p. 55). For these two propositions, and in fact for most other propositions, Gerbillon and Bouvet did add some explanations for them in the lecture notes.

Pyramids and cones are summarily called "pointed solids" (*jianti* 尖體) in the Chinese mathematical vocabulary in late imperial China. In Gerbillon and Bouvet's lecture notes, Article 27 of Volume 5 states:

> 有各種平行底之平面體,各種平面尖體.此底積與彼平行底之平面體底積若等,其高數又若等,則此一平行底之平面體與彼平面尖體三形之積等.再有平行底之圓面體,又有圓面尖體.此底積與彼平行底之圓面體底積若等,其高數又若等,則此一平行底之圓面體與彼圓面尖體三形之積等.[17]
>
> Suppose there are various kinds of planar solids with parallel bases (prisms), and various kinds of [corresponding] planar pointed solids (pyramids). If the base area of this [pointed solid] is equal to that of the [corresponding] planar solid with parallel bases, and the measure

[14]Gerbillon and Bouvet's lecture notes were written both in Manchu and in Chinese, but according to Liu (1991), the structure and contents are essentially identical. So in this paper, only the Chinese version is discussed.

[15]This is a row of smaller characters, indicating that the authors were self-commentating on the title *Jihe yuanben*. Therefore the translation of this row is also written in smaller fonts.

[16]Rare book MS no. 06399, *juan* 1, p. 1, National Central Library, Taipei. Classical Chinese was usually written without punctuations before the twentieth century. All the modern punctuations in the quotations of classical Chinese in this paper are mine, for the convenience of the reader.

[17]Rare book MS no. 06399, *juan* 5, pp. 15–16, National Central Library, Taipei.

of [their] heights are also equal, then [the volume of] this planar solid with parallel bases is equal to the volume of three of that planar pointed solids. Again, suppose there are various kinds of circular solids with parallel bases (cylinders), and various kinds of [corresponding] circular pointed solids (cones). If the base area of this [pointed solid] is equal to that of the [corresponding] circular solid with parallel bases, and the measure of [their] heights are also equal, then [the volume of] this circular solid with parallel bases is equal to the volume of three of that circular pointed solids.

Basically, the statements in this lecture note is not very different from it source in Pardies' text. After the statements and several examples represented in diagrams, Gerbillon and Bouvet ask "Why is it the case?" (*heze* 何則), as in most other propositions in the text. They gave two justifications for this proposition. The first reads:

從一角分為三平面尖體. 此三平面尖體因其底積高度相等, 照前節所云俱為等也. [18]

From one angle [we can] divide [a prism] into three planar pointed solids. Because these three planar pointed solids have equal base and height, according to the previous article, they are all equal.

The strategy they used is to try to convince the reader that one can divide a prism into three pyramids with equal base and equal heights, but this is only possible if the prism is a cube. Thus in this case, Gerbillon and Bouvet are asking the reader to believe that a procedure for the simplest case is good enough for all cases in general.

The second strategy they used is a more "hands-on" approach:

又將各體照實形做空形, 於此空形用水以比例之, 其各體之積自然可得而知矣.[19]

And [we] make empty shapes according to the actual shapes of the solids, and use water to [measure their] proportions. [Then] the volume of each shape can naturally be known.

The term "empty shapes" (*kongxing* 空形) refers to vessels. So, the second method they used is to create vessels and measure their capacity, using real objects to convince the reader about the relations between pyramids and prisms, and between cones and cylinders.

The same article in the *Shuli jingyun* describes the proposition in a virtually identical wording. However, the strategy of demonstration is somewhat different:

如將上下面平行之各體, 以木石為之, 分作同底同高之各平底尖體. 用權衡以較其份量, 則各體之積分自昭然可見矣 (Guo 1993, p. 59).

If [we] take the various solids with parallel upper and lower bases (prisms), and make them with wood or stones. [And we] make respective pointed solids with flat bases and with the same bases and the same heights. Use the scale to measure their weights, and then the volume of each of the solids is naturally elucidated.

As the reader can see, this is again a "hands-on" approach, but this time it does not use vessels and water, but wooden or stone models to demonstrate the proposition to the learner.

[18] *Ibid.*, p. 17.

[19] *Ibid.*

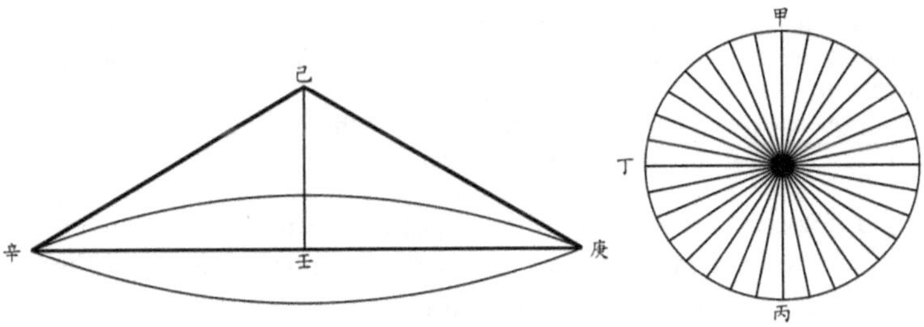

Fig. 1 The sphere and a cone whose volume is equal to the sphere

As for the second group of examples, about the relation between the volume and surface area of a sphere, intuitive argumentations are also used in both the lecture notes and in the compendium.

In Gerbillon and Bouvet's lecture notes, the proposition is stated in what follows:

> 有一球體, 又一尖圓體. 苟尖圓體底面積與球體外面總積若等, 而尖圓體高度與球體半徑又若等, 則此一尖圓體之積為與一球體之積等也. [20]
>
> There is a sphere, and a pointed circular solid (cone). Suppose the base area of this pointed circular solid is equal to the surface area of the sphere, and the height of the pointed circular solid is again equal to the radius of the sphere. Then the volume of this pointed circular solid is equal to the volume of the sphere.

The statement is followed by descriptions of diagrams of a sphere and a cone.

The reader may refer to Fig. 1 (reproduced by the author from the original manuscript). In the lecture notes, *ABCD* is the sphere, with *E* as its centre; *FGH* is the cone with its height *FI*, and base area equal to the surface area of the sphere.[21] After the description of the diagrams, the text starts to demonstrate "why it is the case":

> 將球體從外面至心分為千萬尖體, 此所分千萬尖體之底積必與原球外面之總積等. 然既相等, 即是與己庚辛尖圓體之底面積相等也. 又原尖圓體己壬高度與所分千萬尖體戊甲高度既等, [...] 其己庚辛一尖圓體之積與所分千萬尖體總積等也. 如是其所分千萬尖體之總積既與原球體積之等, 則己庚辛一尖圓體之積必與甲乙丙丁一球體之積等可知矣.[22]
>
> Divide the sphere, from the surface to the centre, into tens of thousands of pointed solids. The [sum of the] bases area of these divided tens of thousands of pointed solids must be equal to the total area of the surface of the original sphere. Since it is so, it is equal to the base area of the pointed circular solid (cone) *FGH*. And since the height *FI* of the original pointed circular solid *FGH* is equal to the height [represented by] *EA* of the divided tens

[20] Rare book MS no. 06399, *juan* 6, p. 81, National Central Library, Taipei.

[21] From the seventeenth to the nineteenth centuries, ten "heavenly stems" and 12 "earthly branches" are used systematically to replace letters in the diagrams in Chinese translations of Western texts. This rule had been practised since Ricci and Xu in their translation of Euclid. See Engelfriet (1998), p. 145. Here in my English translation, I will use A, B, C, D, E, F, G, H, I, J in the places of 甲, 乙, 丙, 丁, 戊, 己, 庚, 辛, 壬, 癸.

[22] Rare book MS no. 06399, *juan* 6, p. 82, National Central Library, Taipei.

> of thousands of pointed solids, [. . .] the volume of the pointed circular solid *FGH* and the total volume of the divided tens of thousands of pointed solids are equal. In this case, since the total volume of the divided tens of thousands of pointed solids is equal to the volume of the original sphere, it can be known that the volume of this one pointed circular solid *FGH* and the volume of [this] one sphere *ABCD* are equal.

So, the strategy of demonstration in this proposition is also an intuitive one, asking the reader to imagine that one can divide the sphere into tens of thousands of tiny cones, and then combine them into a big cone whose base area and height are the surface area and radius of the sphere, respectively.

For the same proposition that was compiled into the *Shuli jinyun*, its wording and strategy of demonstration are essentially identical to those in Gerbillon and Bouvet's lecture notes. So they are not repeated here.

The discussions above have shown that, from Pardies' *Elémens* to Gerbillon and Bouvet's lecture notes to the *Jihe yuanben* in the *Shuli jingyun*, the hypothetico-deductive styles of argumentations in *Euclid's Elements* to prove propositions in a system was not chosen by these French Jesuits and Chinese mathematicians. Instead, a more intuitive approach, together with elementary examples and even hands-on experiments, are used in writing their works. This intuitive style of argumentations, embedded in the imperially composed compendium, influenced not only Chinese mathematicians, but also Korean ones, in the eighteenth and nineteenth centuries.

3 Influences of the *Shuli jingyun* in Korea: The Case of Nam Pyŏng-Gil

Korea's Chosŏn Dynasty inherited much of the mathematical tradition from the previous dynasty, the Koryŏ 高麗 (918–1392). During the time of Koryŏ, Chinese mathematical methods of the tenth to the thirteenth centuries were introduced into Korea through the circulation of new Chinese mathematical texts (Jun 2006). Beside influences from traditional Chinese mathematics in those centuries, there was also indirect influence from Europe in the Chosŏn Dynasty. Korean scholars before the late nineteenth century rarely had the opportunity to study Western learning directly from European scholars or missionaries. Rather, they learned through the Chinese sources incorporating Western knowledge (Kim 1998). During the seventeenth century, several such mathematical treatises were brought to Korea, and most of them were either written by European missionaries or translated from European languages into Chinese. However, the most influential mathematical text in the eighteenth to nineteenth centuries was none other than the *Shuli jingyun*. The earliest record of the *Shuli jingyun* being brought to Korea dates from 1729, only 6 years after its publication in China. This is a result of the efforts of the Chosŏn government and individual scholars to try to keep up with the latest scholarly and intellectual standards (Kim 1998, p. 483).

After its transmission, the *Shuli jingyun* was regarded as an essential reference for Korean scholars in mathematics and astronomy, and it became the most influential mathematical work in Korea in the eighteenth and nineteenth centuries (Oh 2004). In 1791, it was listed as one of the required readings for calendrical astronomy officials (*Chosŏn Wangjo Shillok* 1955, p. 255). By 1818, it had become one of the examination subjects for these officials (Sŏng 1986, p. 336). Therefore, it is not difficult to see that, given the level of importance assigned to it by both the Qing Emperor and the Chosŏn government, those in Korea who wished to study or needed to use mathematics would put the *Shuli jingyun* on their reading list.

The impact of the mathematical compendium from Qing Empire can be exemplified by Nam Pyŏng-Gil's 南秉吉 (1820–1869) commentaries on a Chinese text, the *Jiuzhang suanshu* 九章算術 (Nine Chapters on Mathematical Art, first century CE).[23]

Nam Pyŏng-Gil was from a powerful *yangban* 兩班 family. As the ruling class of Chosŏn Korea, the *yangban* were exempt from the usual service obligations to the state, such as corvée labor and military duty. Devoted to the Confucian doctrine that study and the cultivation of oneself must underlie the governing of others, the privilege that enabled them to become officials took the place of other service obligations to the state. Although ideally the sole profession of the *yangban* was the holding of public office, they seldom occupied the technical posts that were also components of the bureaucracy. Positions for medical officers, translator-interpreters, and specialists who needed to use mathematics—such as accountants or officers in the astronomy and meteorology offices—were all reserved for the *chungin* 中人. Literally "middle people", with a status between the ruling class and the commoners, they were the hereditary class of technical specialists living in the capital (Lee 1984, pp. 172–184, 250–251). Although Nam was from a *yangban* family, he nevertheless had great interests in mathematics. He was one of the most influential mathematicians in his time and wrote at least seven mathematical treatises, including a commentary on the *Jiuzhang suanshu*.

For centuries the *Jiuzhang suanshu* was considered as the most important mathematical canon in China.[24] In Korea, the *Jiuzhang suanshu* was listed as one of the textbooks and examination subjects for mathematics students in the National Academies of different dynasties from the seventh to the fourteenth century (Kim and Kim 1978, pp. 82, 133, 139). However, the text was lost in the middle of the Chosŏn Dynasty before the mid-seventeenth century, and mathematicians stopped discussing the text during China's Ming Dynasty (1368–1644) (Jun 2006, p. 480). In China, the *Jiuzhang suanshu*, along with several other ancient mathematical texts, was recovered in the late eighteenth century.[25] The text was reintroduced to Korea only much later, in the mid-nineteenth century. Shortly after this reintroduction, Nam Pyŏng-Gil published his commentary in 1860s (Jun 2006, pp. 498–499).

[23]For a thorough study of the text, refer to, for instance, Chemla and Guo (2004).

[24]See, for instance, Martzloff (1997), pp. 127–136.

[25]On the recovery of the *Jiuzhang suanshu* in China, see, for example, Li and Du (1987).

Nam explains his reasons for writing this text in the postface of the commentary:

九章算術, 數學之鼻祖也. 劉徽注之, 李淳風釋之, 然俱多未曉處, 抑或繡出鴛鴦而藏其金針之義歟. 注釋所以啓來者, 而終莫能端倪, 故余因原術解之, 發明其萬一, 未敢為覺後覺, 而使好學者庶其易曉云爾 (Kim 1985, p. 495).

The *Jiuzhang suanshu* was the first ancestor of mathematics. Liu Hui commentated on it and Li Chunfeng explained it. However, there are still many places that are not clear. Perhaps [Liu Hui and Li Chunfeng] provided great methods but hid the key to finding their rationale. Commentaries and explanations are for [one] to use to open up future readers' [understanding], but in the end [the previous ones] could not reveal enough clues. Therefore I have explained it according to the original procedures, to bring to light one ten-thousandth of its [meaning]. I dare not [say I can] enlighten those who have not yet understood [the text]. I only hope to make it easier for those who would like to study it.

Nam mentions Liu Hui 劉徽 (fl. c. 263) and Li Chunfeng 李淳風 (602–670) in this postface. Liu Hui is considered by historians of mathematics as the most important commentator on the *Jiuzhang suanshu*, completing his commentary in 263 A.D.; Li Chunfeng was the chief editor of the mathematical canons used as textbooks in the Imperial Academy of his time (Guo and Liu 2001). Obviously Nam does not like the explanations provided by the two authors. I believe that one of the reasons is that mathematicians in his time have been used to the style of argumentations in the *Shuli jingyun*.

In what follows, two examples corresponding to those in the previous section will be provided to show this point.

In the final problem of Chapter 4 of the *Jiuzhang suanshu*, the volume of a sphere is given and the question is to find the diameter. In order to do that, one must know how to calculate the volume of a sphere. Nam does not try to prove the procedure directly from known results in the *Jiuzhang suanshu*. Instead, he quotes two other procedures, which appear in the *Shuli jingyun* in similar forms:

球體外面積應為球徑平圓面積四倍.

外面積與半徑相乘得數, 以三歸之即球積也 (Kim 1985, p. 349).

The surface area of a sphere should be four times the area of the circle [whose diameter] is the diameter of the sphere.

The surface area and the radius multiply each other; the obtained number divided by three is the volume of the sphere.

For this problem, Nam seems to assume that his readers have the knowledge from the *Shuli jinyun*, and directly uses it.

A more interesting case is in Chapter 5 of the *Jiuzhang suanshu*. In this Chapter, two important kinds of solids—the *yangma* 陽馬 and the *bienao* 鱉臑—are discussed. These two polyhedra are shown in Fig. 2: A *yangma* is a pyramid with rectangular base and one lateral edge perpendicular to the base; a *bienao* is another kind of pyramid with right-triangular base and one lateral edge (but not that at the right-angled vertex of the base) perpendicular to the base (Wagner 1979). Their respective volumes are one-third and one-sixth of the product of length, width and height.

Liu Hui's demonstration for the procedures of the volumes of these two polyhedra involves an infinite process (Wagner 1979). Nam did not seem to like it

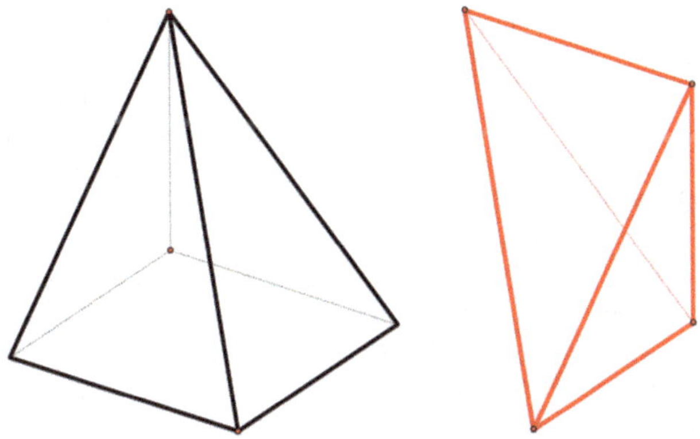

Fig. 2 The *yangma* and the *bienao*

for he never mentioned it at all. The following is Nam's explanation for the volume of a *yangma*:

> 夫一正方體剖之得二塹堵, 一塹堵體剖之得一陽馬一鼈臑, 而一陽馬剖之又得二鼈臑. 是陽馬體為塹堵體三分之二, 即為正方體三分之一. 而鼈臑體為塹堵體三分之一, 即為正方體六分之一也. 乃合二鼈臑體成一陽馬體, 合三陽馬體成一正方體, 故三而一也 (Kim 1985, pp. 362–363).
>
> Now [we have] one cube and cut it open, [and then we] obtain two *qiandu*.[26] [We have] one *qiandu* and cut it open, [and then we] obtain one *yangma* and one *bienao*. And [we take this] one *yangma* and cut it open, [and then] again [we] obtain two *bienao*. This *yangma* solid is two thirds of a *qiandu*, which is one third of a cube. And a *bienao* solid is one third of a *qiandu*, which is one sixth of a cube. So [we] combine two *bienao* solids to one *yangma* solid, and combine three *yangma* solids as one cube. Therefore [for every] three [units, we count] one.

In summary, what Nam says in this piece of commentary is: a cube can be seen as the combination of two *qiandu* (triangular prisms), each of which is in turn the combination of a *yangma* and a *bienao*. But a *yangma* can also be cut into two *bienao*, so one cube is equal to six *bienao* or three *yangma*. Therefore a *yangma* is one third of a cube.

As for the next procedure in the *Jiuzhang suanshu* for calculating the volume of a *bienao*, Nam simply rephrases a part of his argument for the *yangma*, and concludes that a *bienao* is one sixth of a cube.

Mathematically speaking, the *yangma* and *bienao* in Nam's commentary must have equal length, width and height, for they are cut out from a cube. However, in the original problems in the *Jiuzhang suanshu* the length, width and height are all different.[27] Therefore, these two pieces of Nam's commentary ask of the reader that she be able to apply the demonstration of a special case—in fact the simplest

[26] A *qiandu* 塹堵 is a triangular prism whose base is an isosceles right-angled triangle.

[27] Liu Hui used an argument of infinite descent to prove that the ratio between the volumes of a *yangma* and a *bienao* was 2:1 in general situations. A detailed explanation of how Liu Hui proved the procedures for the volumes of the two polyhedra can be found in Wagner (1979).

one—to general cases, appealing to her intuition in order to convince herself that the two original procedures are correct. This method is essentially identical to that for the same problem in the Jesuit's lecture notes. Although it is unlikely that Nam had access to the lecture notes, the idea must have been transmitted through the styles of the *Shuli jingyun*. Therefore, we can see that the intuitive approach that is used in the *Shuli jingyun* is also employed by Nam in this case, among many others.

Nam Pyŏng-Gil was certainly not the only Korean mathematician who employed this method of intuitive approach, nor was he the earliest. Several scholars in the eighteenth and nineteenth centuries, such as Hong Tae-Yong 洪大容 (1737–1783) or Yi Sang-Hyŏk 李尚爀 (1810–late 19th c), also employed a similar approach in reasoning.[28] As far as on what mathematical topics they employed this approach, and on the variations in their reasoning strategies, further investigations are needed in the future.

One might ask why Nam and other mathematicians in the eighteenth and nineteenth centuries would favour such an intuitive approach. As mentioned before, the influence by the imperial compendium was indeed one major factor. However, more can be said in the Korean context. Firstly, Like the Chinese, the Koreans were able to distinguish between the mathematics and the religion which the Jesuits brought and presented together to the East Asian Confucian cultures. Furthermore, the scholars in China and Korea could deconstruct the mathematics into two parts: the "facts"—geometrical propositions and algebraic algorithms—and the reasoning behind them. Confucian scholars in the Chosŏn Kingdom were not particularly fond of the abstract and deductive proofs, because the proofs seemed too "static" for the Neo-Confucian world view. As the Korean Neo-Confucian scholars conceived it, the cosmos was a dynamic network of appropriate interrelations, and any arguments about the "patterns" (*li* 理), including those about geometrical relations, should be addressed with a variety of specific problems and situations. Yi Ik李瀷 (1681–1763), for instance, was a famous scholar who actually read Ricci and Xu's version of the *Elements*, but declined to imitate the deductive approach when he wrote his own mathematical arguments. He also gave numerical examples and calculations when he discussed right-angled triangles.[29]

Secondly, Nam Pyŏng-Gil himself had a good reason to use this intuitive approach. As mentioned earlier, mathematics was essentially an art that was practised by the *chungin* class in the Chosŏn Kingdom. However, Nam had a different view, believing that mathematics should have been included in the education of all Confucian scholars of the ruling class, precisely because it was a necessary tool for ruling and government administration. He wrote several mathematical works, and he published many of his and other mathematicians' works, to try to convince other people in the ruling class that they should study mathematics. In a sense, he was trying to write "popular mathematics" in nineteenth-century Korea (Ying 2014). An

[28] Hong Tae-Yong's and Yi Sang-Hyŏk's mathematical works can both be found in Kim (1985).

[29] A very good discussion about the reactions of the Korean Neo-Confucian scholars to Western mathematics can be found in Baker (2012).

intuitive approach for geometrical argumentations was much easier for many people to understand than a hypothetico-deductive structure.

4 Concluding Remarks

Transmission and interactions among different types of geometrical argumentations constitute some of the most interesting stories in the history of East Asian science and mathematics. At first, Euclid's *Elements*, with its hypothetico-deductive structure, was translated in early 1600s. However, as shown in this paper, it was the intuitive styles of argumentations, easy examples, and even hands-on approaches, partly taken from the popular French textbook *Elémens de Géométrie*, and partly added by Kangxi's court savants Gerbillon and Bouvet, that found a way to convince the Manchu emperor and his Chinese scholars that these kinds of styles made mathematics more intelligible. Therefore, these styles of argumentations were used to compile the geometrical part of Qing Empire's mathematical compendium, the *Shuli jingyun*.

When this compendium was brought to Korea, not only its mathematical methods influenced the Chosŏn Kingdom, but also its styles of argumentations. This paper has presented instances to show that, when the nineteenth-century scholar Nam Pyŏng-Gil was commentating on the *Jiuzhang suanshu*, the paradigmatic text of ancient Chinese mathematics, he also used the methods as well as the styles of argumentations in the *Shuli jingyun*, namely easy examples and the intuitive approach of demonstration.

Scholars in late imperial China and contemporary Korea had access to texts from different traditions and various kinds of styles of geometrical argumentations. How they chose one style over the others is a theme that has not been studied thoroughly, in my opinion, and needs more research in the future so scholars would have better understanding of this interesting topic.

References

Baker, D. (2012). Impotent numbers: Korean Confucian reactions to Jesuit mathematics. *Korean Journal for the History of Science, 34*(2), 227–256.

Burton, L. (1999). Why is intuition so important to mathematicians but missing from mathematics education? *For the Learning of Mathematics, 19*(3), 27–32.

Chemla, K., & Guo, S. (2004). *Les neuf chapîtres: Le classique mathématique de la Chine ancienne et ses commentaries*. Paris: Dunod.

Chosŏn Wangjo Shillok 朝鮮王朝實錄 [The Annals of the Chosŏn Dynasty] (Vol. 46, p. 255). (1955). Seoul: Tongguk Munhwasa.

Chu, P.-Y. (1997). Scientific dispute in the imperial court: The 1664 calendar case. *Chinese Science, 14*, 7–34.

Engelfriet, P. (1998). *Euclid in China: The genesis of the first Chinese translation of Euclid's Elements in 1607 and its reception up to 1723*. Leiden: Brill.

Guo, S. 郭書春 (Ed.). (1993). *Zhongguo Kexue Dianji Tonghui: Shuxue Pian* 中國科學典籍通彙:數學篇 [Sources materials of ancient Chinese science and technology: Mathematics section] (Vol. 3). Zhengzhou: Hebei Education Press.

Guo, S. 郭世荣 (2009). *Zhongguo Shuxue Dianji zai Chaoxian Bandao de Liuchuan yu Yingxiang* 中国数学典籍在朝鲜半岛的流传与影响 [The transmission and influence of Chinese mathematical texts in the Korean peninsula] (pp. 38–46, 74–75). Jinan: Shandong Education Press.

Guo, S. 郭書春, & Liu, D. 劉鈍 (2001). *Suanjing shishu* 算經十書 [Ten mathematical canons]. Taipei: Jiuzhang.

Han, Q. 韓琦 (1993). *Shuli jingyun tiyao* 數理精蘊提要 [Abstracts of the *Shuli jingyun*]. In S. Guo 郭書春 (Ed.), *Zhongguo Kexue Dianji Tonghui: Shuxue Pian* 中國科學典籍通彙:數學篇 [Sources materials of ancient Chinese science and technology: Mathematics section] (Vol. 3). Zhengzhou: Hebei Education Press.

Han, Q. 韓琦 (1995). *Shuxue de chuanru ji chi yingxiang* 數學的傳入及其影響 [The transmission of mathematics and its influences]. In G. Dong 董光璧 (Ed.), *Zhonguo Jinxiandai Kexue Jishushi* 中國近現代科學技術史 [A history of science and technology in early-modern and modern China] (pp. 87–127). Changsha: Hunan Education Press.

Han, Q. 韓琦 (1996). *Junzhu he buyi zhijian: Li Guangdi zai Kangxi shidai de huodong ji chi dui kexue de yingxiang* 君主和布衣之間:李光地在康熙時代的活動及其對科學的影響 [Between the Monarch and a civilian: Li Gunagdi's activities in Kangxi's time and its influences to science]. *Tsing Hua Xuebao* 清華學報, *26*(4), 421–445.

Han, Q. 韓琦 (1999). *Gewu qiongli yuan yu mengyiangzhai – shiqi shiba shiji zhi zhongfa kexue jiaoliu* 格物窮理院與蒙養齋— 17、18世紀之中法科學交流 [*Gewu qiongli yuan* and *Mengyiangzhai* – Sino-French scientific exchanges during the 17th and 18th centuries]. *Faguo Hanxue* 法國漢學 [Sinologie française], *4*, 302–324.

Han, Q. 韓琦 (2003). *Kangxi shidai de shixue jiaoyu ji chi shehui Beijing* 康熙代的教育及其社背景 (Mathematics education during Kangxi's time and its social backgrounds). *Faguo Hanxue* 法國漢學 [Sinologie française], *8*, 434–448.

Han, Q. 韓琦 (2008). *Yesu huishi he Kangxi shidai lisuan zhishi de chuanru* 耶穌會士和康熙時代曆算知識的傳入 [Jesuits and the transmission of knowledge about calendrical astronomy during Kangxi's time]. *Aomenshi Xinbian* 澳門史新編 [A new edition of the history of Macao] (pp. 967–986). Macao: Fundação Macau.

Horiuchi, A. (2010). *Japanese mathematics in the Edo period (1600–1868)* (pp. 6–7). Basel: Birkhäuser.

Horng, W.-S. (1993). Hua Hengfang (1833–1902) and his notebook on learning mathematics – *Xue Suan Bi Tan. Philosophy and the History of Science: A Taiwanese Journal, 2*, 27–76.

Huang, Y.-L. (1991). Court divination and Christianity in the K'ang-hsi era. *Chinese Science, 10*, 1–20.

Jami, C. (1996). From Clavius to Pardies: The geometry transmitted to China by Jesuits (1607–1723). In F. Masini (Ed.), *Western humanistic culture presented to China by Jesuit missionaries (XVII–XVIII centuries)*. Rome: Institutum Historicum S. I.

Jami, C. (2004). Teachers of mathematics in China: The Jesuits and their textbooks (1580–1723). In L. Saraiva (Ed.), *History of mathematical sciences: Portugal and East Asia II* (pp. 92–93). Singapore: World Scientific Publishing.

Jami, C. (2012). *The emperor's new mathematics: Western learning and imperial authority during the Kangxi reign (1662–1722)*. New York: Oxford University Press.

Jami, C., & Han, Q. (2003). The reconstruction of imperial mathematics in China during the Kangxi reign (1662–1722). *Early Science and Medicine, 8*(2), 88–110.

Jun, Y. H. (2006). Mathematics in context: A case in early nineteenth-century Korea. *Science in Context, 19*(4), 475–512.

Karpinski, L. C., & Kokomoor, F. W. (1928). The teaching of elementary geometry in the seventeenth century. *Isis, 10*(1), 21–32.

Kim, Y. S. (1998). Problems and possibilities in the study of the history of Korean science. *Osiris, 13*, 48–79.

Kim, Y.-U. 金容雲 (Ed.). (1985). *Hanguk Kwahak Kisul Sa Charyo Taekye: Suhak P'yŏn* 韓國科學技術史資料大系:學篇 [Source materials collection in the Korean history of science and technology: Mathematics section] (Vol. 6, p. 495). Seoul: Yŏgang Press.

Kim, Y.-U. 金容雲, & Kim, Y.-G. 金容局 (1978). *Kangoku Sugakushi* 韓国数学史 [A history of Korean mathematics] (pp. 82, 133, 139). Tokyo: Maki Shoten.

Kokomoor, F. W. (1928). The teaching of elementary geometry in the seventeenth century. *Isis, 11*(1), 85–110.

Lee, K.-B. (1984). *A new history of Korea* (pp. 172–184, 250–251). Cambridge, MA: Harvard University Press.

Li, Y., & Du, S. (1987). *Chinese mathematics: A concise history* (J. N. Crossley & A. W. C. Lun, Trans.). Oxford: Clarendon.

Liu, D. 劉鈍 (1991). *Shili jinyun zhong jihe yuanben de diben wenti* 數理精蘊中幾何原本的底本問題 [The problem of the basis for the *Elements of Geometry* in the *Shuli jingyun*]. *Zhongguo Keji Shiliao* 中國科技史料 [Historical Materials of Chinese Science and Technology], *3*, 88–96.

Martzloff, J.-C. (1997). *A history of Chinese mathematics* (pp. 127–136). Berlin: Springer.

Mei, R. 梅榮照, Wang, Y. 王渝生, & Liu, D. 劉鈍 (1990). *Oujilide yenben de chuanru he dui woguo ming qing shuxue de yingxiang* 歐幾里德原本的傳入和對我國明清數學的影響 [The transmission of Euclid's *Elements* and its influence to mathematics of the Ming-Qing period of our country]. In R. Mei 梅榮照 (Ed.), *Ming-Qing Shuxueshi Lunwenji* 明清數學史論文集 [A collection of papers about the history of mathematics in Ming and Qing] (pp. 53–83). Nanjing: Jiansu Education Press.

Oh, Y. S. (2004). *Suri chŏngon pohae* (數理精蘊補解): An 18th century Korean supplement to *Shuli jingyun* (數理精蘊). In W.-S. Horng, Y.-C. Lin, T.-C. Ning, & T.-Y. Tso (Eds.), *Proceedings of Asia-Pacific HPM 2004 conference*. Taichung: National Taichung Teachers College.

Pardies, I. G. (1673). *Elémens de Géométrie, ou par une methode courte & aisée l'on peut apprendre ce qu'il faut sçavoir d'Euclide, d'Archideme, d'Apollonius, & les plus belles inventions des anciens & nouveaux Geometres*. Paris: S. Mabre-Cramoisy.

Parsons, C. (1979). Mathematical intuition. *Proceedings of the Aristotelian Society NS, 80*, 144–168.

Shapiro, S. (2000). *Thinking about mathematics* (pp. 202–211). New York: Oxford University Press.

Sŏng, C.-T. 成周悳 (1986). *Sŏungwan Chi* 書雲觀志 [Record of Sŏungwan]. In K.-N. Yu 俞景老 (Ed.), *Hanguk Kwahak Kisul Sa Charyo Taekye: Ch'ŏnmunhak P'yŏn* 韓國科學技術史資料大系:天文學篇 [Source materials collection in the Korean history of science and technology: Astronomy section] (p. 336). Seoul: Yŏgang Press.

Spence, J. (1984). *The memory palace of Matteo Ricci*. New York: Vintage.

Tian, M. 田淼 (2003). *Zhongguo Shuxue de Xihua Licheng* 中國數學的西化歷程 [The westernisation of mathematics in China] (pp. 26–38). Jinan: Shandong Education Press.

Wagner, D. B. (1979). An early Chinese derivation of the volume of a pyramid. *Historia Mathematica, 6*(2), 164–188.

Ying, J.-M. (2014). Mathematical canons in practice: The case of a nineteenth-century Korean scholar Nam Pyŏng-Gil and his evaluation of two major algebraic methods used in East Asia. *East Asian Science, Technology and Society: An International Journal, 8*(3), 347–362.

FSC
www.fsc.org
MIX
Papier | Fördert
gute Waldnutzung
FSC® C083411

Kolibri 360 GmbH
Ferdinand-Jühlke-Straße 7
99095 Erfurt, Deutschland
produktsicherheit@kolibri360.de